THE CHANNEL

THE CHANNEL

The Remarkable Men and Women Who Made It the Most Fascinating Waterway in the World

CHARLIE CONNELLY

WEIDENFELD & NICOLSON

First published in Great Britain in 2020 by Weidenfeld & Nicolson
an imprint of The Orion Publishing Group Ltd
Carmelite House, 50 Victoria Embankment
London EC4Y 0DZ

An Hachette UK Company

1 3 5 7 9 10 8 6 4 2

A CIP catalogue record for this book is
available from the British Library.

ISBN HB 978 1 4746 0791 9
ISBN eBook 978 1 4746 0793 3

Typeset by Input Data Services Ltd, Somerset

Printed and bound in Great Britain by Clays Ltd, Elcograf S.p.A.

MIX
Paper from
responsible sources
FSC

Contents

1

Sound Mirrors

They look almost otherworldly, the old sound mirrors at Denge just off the road to Dungeness.

There's always a bleak feeling when you come off the marsh onto that giant spit of shingle. It's the Channel at its rawest and most elemental, somehow, washing up against land so low and flat you suspect the sea could come rolling in and cover it on a whim as if it was never there. The air feels thick, like breathing ether, and time seems to work in a different way: there are some parts of Dungeness where ancient events feel like they've just happened, you can sense them like a presence, like going into an empty room that was full of people a moment ago. It's not about ghosts, just a compression of time.

After watching the breakers rolling in from a long way out for a while, I turn and walk away from the Channel up a street of bungalows to an expanse of reedy, grassy shingle. The sound of the sea vanishes and on this warm day the air here feels even thicker, like soup. I head north and catch a distant glimpse between the trees that surround them on their little island home in the flooded gravel pits. In the far distance, four miles away or more, are the lighthouses at the tip of Dungeness and the grey box silhouettes of the nuclear power station, but I'm walking away from the these and heading for something few people visit and which by rights shouldn't really be there any more.

They grow bigger in the glimpses as I draw nearer, two giant scallops and a long concave wall, until I can walk along the path through a copse and suddenly they're right in front of me, three enormous and eerie concrete structures facing the sea.

These are the old ears of the English Channel, radar before there was radar. Built in the late Twenties and early Thirties, the two giant concave cups and long, curved wall in front of me were designed to pick up the sound of approaching aircraft or artillery, their scooped shape cast specifically to focus sound towards a single point where a microphone would relay it to operators housed in the belly of the structure. At their most advanced they could give a fifteen-minute warning of imminent air attack.

There's still a noble dignity about the sound mirrors, even though they were obsolete almost as soon as they were built and the concrete edges of their cups have been crumbled by decades of wind. They're in the middle of a nature reserve now and can only be visited on open days a couple of times a year, but you can still get close. I stand dead still and listen to the rustle of the breeze, the cluck and chirrup of birds in the reeds and every now and then the throaty mosquito whine of a small aircraft taking off from the airfield in the distance across the marsh at Lydd. Beneath the ambient noise I'm sure I can hear something else. I try to listen but it's just out of reach, a sense that something's about to happen while at the same time it happened a very long time ago indeed.

I look across the water to the mirrors and I realise what it is I can almost detect: the echoes of stories. The sound mirrors are redundant, their microphones long gone, but they're still eavesdropping on the waves, still drawing in sounds, collecting the faint echoes of the past that always gather at Dungeness.

They were positioned to watch the sea but in the intervening years the houses of Greatstone have inched along the front. Today there are roofs, windows, chimneys and satellite dishes between the mirrors and the horizon, but while they can't see it any more, and while they might be relics now, they're still listening to that horizon. Tilted slightly to the sky at Easter Island angles, they're still gathering in the sounds of the birds, the reeds, the wind, the aircraft, and most of all they're still listening to the English Channel, still picking up its stories.

2

The Pride of Britain

I'm writing this roughly halfway across the English Channel. It's a sunny, hazy morning as I look out across the prow of the ship from the forward lounge. This isn't a busy crossing but there's a lively hubbub of conversation, mainly from a large group of English schoolchildren excited by the change of routine, with no books, no lessons, just the prospect of whatever European journey lies ahead. I'm just back from the bar where I've purchased a Starbuck's coffee and croissant, roughly the equivalent of stopping for a petrol station pork pie on the way to Melton Mowbray, but patisserie options are limited here in the middle of the Channel.

In the queue, waiting to be served, I stood behind a short man with grey hair cropped in an almost military style. He had one of those remarkable stomachs that are almost perfectly round and start from somewhere in the middle of the chest. It looked firm enough to deflect a bullet. He also appeared to be operating entirely from within a cloud of last night's booze, that sharp, almost metallic fragrance that indicates alcohol is being expelled from each and every pore. It was a tang so overwhelming I feared that if he went anywhere near a naked flame he might actually go up.

The man behind the bar spoke no French, the human sphere in front of me no English, so he gestured in the manner of a man playing a vertical concertina that it was a large beer he required. The barman waggled a pint glass in the air and received thumbs up in return. The man on the other side of him, of strikingly similar age, build and hairstyle, and wearing a Help For Heroes hoodie, copped a noseful of stale booze, looked round and regarded his French near-doppelgänger disapprovingly over the top of his reading glasses, a look so sharp I'm sure I saw the cheese curdle in the toasted sandwich he'd just collected.

I manoeuvred the coffee and croissant back to my seat at a table next to two experienced cross-Channel travellers, a pair of middle-aged women sipping tea dispensed from a flask beneath formidable hairstyles that looked as if they'd not only remain undisturbed by a tornado but would also provide me with an effective shelter, should I climb inside one of them. Their ferry routine was so slick and honed that I knew they were at the best table and pulling the little red tag on their packet of Hobnobs within about ninety seconds of the car ramp clanking down onto the tarmac at Dover.

The sun glints on the sea ahead through salt-encrusted floor-to-ceiling windows that lend a gauze-like quality to the view. I close my eyes and immerse myself in the hubbub of conversation. There's almost a purgatory feel, a sense of calm fringed with slight impatience. For all the efforts to make the lounge into a place you'd want to come and hang out even if we weren't stuck in a big iron canister on the sea, nobody's truly settled here. We're all slightly fidgety, anxious to continue our journeys, not quite resentful of having to abandon our cars inside a floating metal car park for a couple of hours but with the air that we could do without it. Conversations are distracted and peter out in the shifting hubbub, heads turning to look out at the sea or at the television on the wall soundlessly playing a French news channel.

'Just because you *can* enter into a sexual relationship, Ethan,' crackles a recently broken teenage male voice above the ambience, 'doesn't mean you *have* to.'

The MS *Spirit of Britain* is, along with her sister ship *Spirit of France*, the largest passenger vessel ever to ply the Dover–Calais route. She has capacity for 1,059 cars and 180 lorries accommodated in 2,700 metres of lanes. The ship can carry 2,000 people, weighs a shade under 50,000 tons and is twice as long as the frontage of Buckingham Palace. Introduced in 2011, this vessel is at the zeitgeist of Dover–Calais passenger shipping. There's barely a sense you're at sea beyond the signs for muster stations and the engines rumbling away in the bowels of the ship nine decks below the vibrating meniscus of my coffee.

Earlier that morning I'd swooped down from the top of the White Cliffs on the elevated road that curves above the Port of Dover as distant ferries disappeared noiselessly into the haze towards Dunkirk and Calais. I'd then swept through Dover port flanked on both sides by empty lanes and barely needing to stop. A quick check of my passport first by a cheery Englishwoman then a bored-looking Frenchman and I was able to drive onto the ship almost without delay, disappearing into its huge metallic maw before being pointed up a slope to a higher level, where I chugged slowly forward under instruction from the beckoning fingers of a man in fluorescently grubby overalls staring down at the reducing gap between mine and the car in front. With a flat sea and barely a breath of wind beneath the blazing early morning sunshine, this is about as good as a crossing of the English Channel has ever got.

Fortunately, most of us are too young these days to remember the times when cross-Channel passenger ships were small enough to respond to every wave with a pitch, roll or full-on rivet-jiggling judder. In bad weather the crossings could take up to eighteen hours, after which green-faced passengers would stagger off the ship, still retching and swooning, onto a crowded quayside before having to pass through a cavernous, draughty customs shed. Even on less stormy days passengers tended to elbow each other out of the way to find a place to sit by something they could hang on to that was bolted down. Believe me, the last thing on Ethan's mind would have been getting laid.

In 1924, for example, the artist Estella Canziani crossed the Channel to Dover after a summer spent painting in France. It was a windy day and she recalled 'climbing up a tank on the deck to escape the waves breaking over the ship'. The gangways below were 'crammed with ill passengers' and she feared for the summer's work she had in a box beside her at the expense of her outfit. 'Clothes do not matter; they will wash and dry,' she wrote, 'it was like taking them out of a pail of water.'

'The grumbles begin at the start of the boat train, increase on embarkation, reach their maximum in the Channel, continue in another form on disembarkation and do not subside until the journey is over,' was *The Times*'s summation of the nineteenth-century cross-Channel experience, while even for habitual Channel criss-crosser Charles Dickens, writing in *The Uncommercial Traveller* during the early 1860s, the journey was rarely less than hellish.

'A stout wooden wedge driven in at my right temple and out at my left, a floating deposit of lukewarm oil in my throat, and a compression of the bridge of my nose in a blunt pair of pincers,' he wrote, 'these are the personal sensations by which I know we are off, and by which I shall continue to know it until I am on the soil of France.'

Even a glimpse of journey's end wasn't much consolation.

'Malignant Calais!' he wailed. 'Low-lying alligator, evading the eyesight and discouraging hope! Dodging flat streak, now on this bow, now on that, now anywhere, now everywhere, now nowhere!'

During the 1920s the barrister and Conservative MP Sir

Ellis Hume-Williams set down his experiences of the process through which I had just passed. On checking in, he wrote, passengers were required to haul their baggage to a window where they heaved it onto the ledge and then let it fall – to be caught, 'more or less', by a porter, who then hauled it away while the passengers milled around in pens waiting for their passports to be checked before boarding. Once on the ship, things were no less chaotic.

'There were no seats for a third of us,' Ellis-Williams grumbled. 'People were sitting and lying on the decks, on the floors of the cabins, anywhere.'

As the ship neared its destination the crew began a little freelance opportunism, touting loudly for porter fees. Once ashore, Hume-Williams was approached by a more official-looking porter 'who hurriedly pressed into my hand a dirty piece of paper inscribed with the number 162 and left'.

Inside the customs shed, a pandemonium of porters and passengers trying to find each other, Hume-Williams was unable to discern among the clamour anyone shouting his number, so began shouting it himself, a method then taken up by several other travellers. 'Personally I was more lucky than most of the choir because, moving forward to get a better position for my voice, I fell over my own suitcase which my porter had thoughtfully left in the middle of the floor before deserting me for a more opulent client.'

A century earlier, in August 1814, Charles Farley, actor, and the manager of the Covent Garden Theatre, left this account of his return from Boulogne to Dover in the company of the composer Henry Bishop:

At four o'clock embarked on board the Industry for Dover [he wrote in his diary]. Over five hours beating about the French coast. Rain came on, very dark, vessel tossed about very much.

Giddy and could not be sick. Bishop very ill. Breeze came on and we stood tight before the wind and in near three hours finally reached the Strait of Dover. Could not enter the harbour, came to anchor, tossed about, rain very heavy. Hailed a boat, came alongside, much difficulty getting in. Tossed about as not enough water in the harbour for even a boat to enter. Carried on the backs of sailors. Went to the King's Head, supper on Pidgeon Pye. To bed.

Things had been pretty similar in the autumn of 1776 when a twenty-one-year-old woman from Rugeley in Staffordshire named Mary Capper embarked for Boulogne on the *Four Friends* from the Thames by the Tower of London, on a journey designed to aid her ailing health. With the American War of Independence taking place across the Atlantic, the captain couldn't even tell his passengers when they would depart. The government, urgently needing recruits for the Navy, had issued a torrent of press warrants to which the skipper of the *Four Friends* kept losing crew members. A combination of the press gangs and the fog kept the ship from departing for fully nine days after Capper had arrived at the original boarding time. When the ship could finally leave, conditions weren't ideal.

'The cabin was exceedingly crowded so that it was by no means agreeable when one considers it was the State Cabin where seventeen of us were to eat, drink and sleep,' she wrote in her journal. 'However as I knew that complaints would be to no avail I sat down in a snug corner and resolved to make things as agreeable as possible. Nothing could exceed our embarrassment as the hour of rest approached but the captain, guessing our wishes, obligingly took the gentlemen up on deck.'

It was a dreadful crossing that ended up taking a full three days, including a hair-raising night spent sheltering in the Downs – the stretch of water between the Goodwin Sands and the Kent coast that provides a modicum of protection from stormy seas – while expecting another ship to come crashing into them at any moment during the storm-tossed night. Finally they made it to Boulogne on 11 November, 'and never did any poor creatures receive their liberty with greater joy'.

Rupert Brooke recorded his experience of a rough crossing in his 1909 poem 'A Channel Passage', in which he compared the nausea of a stormy voyage to France to his troubled love life. 'The damned ship lurched and slithered,' he wrote. 'Quiet and quick my cold gorge rose; the long sea rolled; I knew I must think hard of something or be sick; And could think hard of only one thing – you!'

While Ethan might have approved, this was about as far from the sun-drenched, honey-for-tea homesickness-inspired fuzz of 'The Old Vicarage, Grantchester' as you could imagine, as 'retchings twist and tie me, old meat, good meals, brown gobbets up I throw'.

Fifty years earlier Algernon Charles Swinburne had had a similar experience, one that he turned into his poem 'A Channel Passage'. Having steamed out of Calais in fair weather, soon 'such glory, such terror, such passion as lighten and harrow the far, fierce East; rang, shone, spake, shuddered around us: the night was an altar with death for a priest'.

You can even go back as far as the twelfth century to find similar evidence of the Channel's hold over our minds and our stomachs. That was when King John granted one Solomon Attfield land at Dover on condition that 'as often as our Lord the King should cross the sea the said Solomon or his heirs should go along with him to hold his gracious head on the sea, if it was needful'.

Considering these were people crossing the Channel by choice at its narrowest point, is it any wonder that the Channel has been such a bulwark? For centuries this narrow stretch of sea has been an impenetrable barrier guarding us against invasion, keeping us separate from the rest of the world. Its presence has been reassuring; for some it's permitted the nurture of an ostensibly proud but parochial sense of nationality: the famous headline 'Fog in Channel, Continent Cut Off' is apocryphal but endures because its expression of our island mentality is entirely plausible, something that has come into much sharper focus since the vote to leave the European Union made the Channel, which we had become used to as a bridge, into something more like a moat.

The symbolism of the Channel goes way beyond a simple role as a conduit for shipping, trawling and the odd lard- and lanolin-slathered maniac in goggles and swimming trunks plunging in and making for the other side. It's made us believe we're more separate than we are thanks to 'that strip of sea which severs merry England from the tardy realms of Europe', as the *Church and State Review* put it in 1863.

Yet that severance isn't as wide as you'd think. I live in Deal on the Kent coast and on a clear day from my garden I can see France. At night the lights of Calais, Gravelines and Dunkirk throw an orange glow into the sky clearly visible from my home: continental Europe is so close you can see its streetlights, and I'm not even at the Channel's narrowest point. The guns of the Western Front could be heard along the English coast; the Hawthorn Ridge Redoubt explosion, heralding the start of the Battle of the Somme, was heard in London. Spike Milligan, stationed in a concrete pillbox at Bexhill-on-Sea, clearly heard the thuds and booms of the Normandy landings. The Channel kept the danger at bay but it was closer than we liked to believe.

In the late summer of 1802 William Wordsworth arrived back in England from a visit to France. The Treaty of Amiens, signed that spring, had provided a rare rapprochement in Anglo-French hostilities that would last for a year, known also as the

Peace of Amiens – meaning the English Channel that summer brimmed with travellers criss-crossing like never before. As Wordsworth made his way north from Dover he paused for a moment and looked back towards the sea, and wrote:

> Inland, within a hollow Vale, I stood,
> And saw, while sea was calm and air was clear,
> The Coast of France!
> The Coast of France, how near!

Considering he'd just arrived from the coast of France, this shouldn't have come as an enormous shock to Wordsworth. If he'd found himself regarding the Alps from the edge of the South Downs it might have warranted the repeated exclamation, but the poet's realisation that our nearest continental neighbour is close enough to be clearly visible from our own territory serves as a neat summary of our general misapprehension of the English Channel.

I'm probably being a little unfair to Wordsworth, as the point he was making was as much political as geographical – although given he'd expressed just a few weeks earlier similar thunderstruck surprise at some daffodils, he was sailing uncomfortably close to being a bit of an old gawd-help-us – but exclamations like his exemplify the strange combination of pride and indifference that characterises our relationship with the Channel.

Advances in technology, transport and human endurance have narrowed the Channel gap even further. Crossings have become easier, more varied and more regular since the early days of sail and steamer voyages. Nowadays when we take the Eurostar to France it goes dark for twenty minutes or so as we pass between the nations, and that's it, we barely look up from our newspapers or the *Great British Bake Off* on the iPad. If we take the Eurotunnel we don't even have to get out of the car before we're in France – that's in less time than it takes to read the whole safety notice on the wall of the carriage and get through a whole bag of Haribos. Modern cross-Channel ferries like the *Spirit of Britain* are the dimensions of a medium-sized industrial estate, and are large enough to ensure most journeys are no more uncomfortable than negotiating the high street on a skateboard, rendering accounts like those of Hume-Williams, Brooke and Swinburne no more than curiosities to us now. Advances in Channel transportation have made crossing the Channel so easy that sometimes we barely even notice we've done it.

Rather than the crash of the waves over the bows and a chorus of retching, sitting in the lounge with my coffee and croissant I'm sensing little more evidence of motion than the distant clink of bottles in the duty-free shop keeping perfect time with the throb of the engines. The only significant danger I can identify is the flammable Frenchman sitting a few tables away nursing his *poils du chien*.

I can't see Calais through the haze this morning, but I know it's there. I'm not sensing any of that Dickensian malignance, either. Indeed, I'm looking forward not only to seeing Calais but exploring it. Unlike just about everyone else on the *Spirit of Britain* I'll be following signs from the port not for the motorway south but for Calais-ville. Since I moved to live at the mouth of the English Channel I've become fascinated by its people, places and stories. So much so that I've even started regularly throwing myself into it.

3

Swimming in the Channel

The United Airlines flight from Chicago to Frankfurt is flying at a shade under 37,000 feet when it crosses the English Channel just as dawn breaks on the day I head over to Calais. It's been in the air for seven hours and has about one more to go until it lands. The plane is travelling at 525 knots, a little over 600 mph, and right now the cabin crew are probably serving breakfast to bleary travellers rubbing an eye with one hand and lowering their tray-table with the other. Foil lids are peeled back to reveal rubbery scrambled egg, and everyone tries to keep their elbows tucked in. Most of the window blinds are still down from the overnight flight, so it's fair to say that not many people on board will have noticed the silver arrowhead of water passing below them in the yellow-grey light of dawn.

I know it's a United Airlines flight from Chicago to Frankfurt because I'm standing beneath it holding my phone up to the sky and looking through an app that tells me that's exactly what it is. It's the highest of around half a dozen white darts above me, wakes on the sky-sea, all of them so high as to be completely silent, heading towards the yellow light glowing from the eastern horizon.

On my screen the tags over the little aeroplane symbols tell me there are also flights passing over me from Stansted to Rome, from Birmingham to Munich and Atlanta to Stuttgart. I'm watching them from the very edge of the Channel, so close to the water the waves are lapping over my feet. The sun is yellowing the horizon and throwing up a blush of pink on the whispers of cirrus above, while behind me to the right hangs the half-moon in a semi-darkened sky. Watching the distant specks with their short white trails I calculate that crossing the English Channel here, close to its narrowest point, takes about two minutes at 600mph. The United flight had just crossed the Channel in less time than it took Irv Schwartz from Wisconsin, wedged into a middle seat and on his way to a bathroom-fittings expo in Mainz, to shred his crumbly breakfast roll with the rock hard slab of butter stuck to the end of his plastic knife.

For a stretch of water that has helped to define us politically, culturally and economically we seem to take little notice of the Channel these days. For most of us it's an inconvenience to be traversed on the way to the rest of Europe and beyond that we only notice if it tugs at our sleeve through bad weather or rough seas that might have us doing what I've just christened 'the Rupert Brooke yodel'. We don't see one of the most important stretches of water in the world that has facilitated some of history's greatest journeys of exploration, from Francis Drake to the Pilgrim Fathers to Captain Cook to Franklin's doomed expedition and to the 1953 Everest team. We don't see the busiest shipping lane in the world where more than 500 vessels a day pass through: bulk carriers, naval vessels, tankers, container ships, ferries, dredgers, lighters, coasters, cruise ships, trawlers, coal barges and yachts – the full range of maritime hardware sailing by all day, every day, all year round. We don't see a submerged landscape of shifting sandbanks, tides and eddies, cables and tunnels, shoals of fish and plant life and countless centuries of shipwrecks that litter its floor. If we see anything at all it's a featureless expanse of water when that is the last thing the English Channel could ever be.

For one thing, by a rough estimate there could be as many

as 20,000 people in transit in the Channel right now as you read this: cargo ship crews, ferry passengers, people on yachts and fishing boats, and that's before thinking about the Eurostar and Eurotunnel passengers passing beneath. There are people making journeys out there for all sorts of reasons: their living, on business, on holiday, or just experiencing the sheer pleasure of being on the water. There are also the coastguards and lifeboat crews, the port and dock workers and the naval personnel on both sides of the Channel, ensuring it remains a living, thriving entity rather than the featureless slice of blue that appears on maps.

Also out there are centuries of history, of invasion, exploration, tragedy, disaster and triumph. From the Romans to the Normandy landings the Channel has witnessed some of the world's most famous and significant military encounters. It's witnessed pioneering journeys and courageous feats of endurance that became landmarks in the history of human and technological development: Jean-Pierre Blanchard's first crossing of the Channel by balloon in 1785, Blériot's by plane in 1909 and Captain Matthew Webb becoming the first person to swim the Channel in 1875 are all immense landmarks in the history of human development. It was all very well going up in a balloon or flying an aeroplane, but until you'd crossed the Channel it didn't really mean zip.

The Channel was the ultimate test of endurance, heroism and progress.

Coming from long lines of ancestors connected with the sea – I'm descended from a heady briny mixture of mariners, shipwrights and dockers – I've always been excited by the prospect of a vast expanse of water stretching to the horizon because

I know that just beneath that rippling surface lies an endless supply of stories. Wherever you look, whichever sea is laid out in front of you, there are centuries' worth of them: tales of heroism, disaster, romance, discovery, mystery and ghosts; and the English Channel – and the eastern Channel in front of me in particular – is supplied with more of them per square yard than any sea you'd care to mention. Which partly explains why, once I'd followed the United flight to Stuttgart until it vanished into the haze, I put my phone down on the beach, walked into the water until it was up to my neck, put my arms out in front of me and struck out towards the horizon.

I've never been a particularly gifted swimmer and I was certainly never an enthusiastic one until I moved to within a stone's throw of the Channel. Before then, if you'd asked me to make a list of things I'd quite like to do, it's safe to say swimming in the cold, choppy waters of the English Channel would have ranked well below going to the dentist and calling the customer-service line of a utility company.

I'm not even sure what prompted me in the first summer I lived here to buy a pair of bright-red swimming shorts, put them on, crunch purposefully over the shingle, put one foot into the freezing, foamy fringe of a wave sloshing in over the stones, yelp, swear, turn around and crunch purposefully home again. But even though the start was inauspicious it wasn't, by any means, the end of it. Ordinarily that would have been that, the swimming shorts would have gone to the back of the drawer never to be spoken of again, and my physical regime would have defaulted to carrying the bins out once a week and opening bags of crisps with a grunt. This time, however, I persevered. I stuck at it. I tried the Channel again, wading in up to my shoulders, albeit so slowly the tide came in and out twice around me, and kept returning until eventually I found myself trotting down to the sea first thing every morning all summer.

I even bought some special gear, the first items of sportswear I'd bought that weren't Charlton Athletic replica shirts since my teens: a 'wearable towel' (a big dress with a hood) for the summer and a 'sports cloak' (a big dress with a hood and a fleece lining) for the colder mornings.

I'm addicted now. I wake up ridiculously early to catch the shipping forecast at twenty past five, listening out for sea area Dover then waiting for the inshore waters forecast for North Foreland to Selsey Bill. Carrying my thermal mug of tea I head out into the dawn, catching my breath every morning at my first sight of the English Channel, and then I'm making my way across the shingle and descending towards it. The steep angle of the beach down to the sea means that by the time I reach the water the town has disappeared behind the stones and it's just me, the beach, the sky and the Channel. There's nobody else down there at that time, the entire sweep of the beach is all mine as far as the eye can see in both directions. The sensation of the cold water makes every part of me thrum with energy, every sinew, every nerve ending practically singing, and I've never been more aware of my own body.

Eventually my feet leave the bottom and I'm weightless and swimming. If I've timed it right the sun rises over the horizon while I'm in the water, a glint at first, a sparkle, then a perfect rising semicircle of burning orange that takes on just for a moment the shape of a hot-air balloon as the Channel horizon clings onto the lifting ball before the sun detaches, scattering shifting orange shards across the surface of the water. There are more exotic places than this to watch the sun rise, but take it from me, on such mornings the English Channel is the most beautiful thing in the world.

There are of course days when it's nothing like that, when the mornings are grey and dark and the Channel is grey and dark and there are bullets of stinging rain in the wind and the waves are relentless and malevolent and the last thing I feel like doing is stripping off and sticking my face into the freezing, stroppy brine in front of me. The more I did it, though, the more I swam in these waters where the North Sea ends and the Channel begins, the more connected I felt to the stretch of water filling the shallow crevice that connects Britain to the rest of Europe. Soon it was about far more than exercise, more than a bit of fun, it was a daily routine that I relished, one that's always the same yet always different: a different sky, a different tide, a different current, a different colour sea from grey to brown to green to deep blue to Caribbean turquoise.

When I come out of the water after half an hour or so I'm overcome by a mixture of physical fatigue from the exertion and an all-encompassing sense of calm. I'll stand at the water's edge sipping tea with my hood up in the morning mist like some kind of spectral maritime monk and I'll watch the rising sun, or the red buoy bobbing halfway to the horizon, or the giant container ships drifting in silhouette along the Channel meniscus like distant cities, or the ferries easing in and out of Dover docks with their lights pin-sharp in the dawn gloaming.

I'm not always alone: there have been mornings when I've come face to whiskery face with a seal and I'm regularly butted amidships by what feel like pretty decent-sized fish. One morning I found myself swimming among a shoal of mackerel, which was a little like being assailed by underwater hailstones. Local fishing boats sometimes chug past me, or the occasional rowing eight from the club on the other side of the pier. Some mornings I'll see the faintly sinister presence of one of the battleship-grey Border Force cutters just offshore and when bad weather's forecast there'll often be some of the smaller container ships or tankers anchored just off the coast, taking advantage of the protection of the Downs until the danger's passed. On quiet mornings I can even hear faint hints of the crews' conversations on the breeze.

The more time I was spending in the Channel, the more attached to it I was growing. At first it was just about my daily immersions, my addiction to the burning on my skin induced by the cold water and the sloshing sound of the water around my ears as I swam. My last thought at night was of the next morning's swim and my first thought in the morning was of the weather and the sea conditions. Then I found myself thinking more and more about the Channel in a wider context. As well as the fact that it had forged the daily news cycle almost every day since the vote to leave the European Union, it was a constant physical presence. I could see it from the garden; even when I wasn't swimming I was spending my days and nights a matter of yards away from it. When the windows were open I could hear it and sometimes even smell it inside my own home.

In and out of the water I found myself thinking about the Channel's place in our history and national psyche, and its influence on both, positive and negative. I'd think about the various crossings it's inspired over the centuries, from Julius Caesar's to the deflating dinghies for which half a dozen freezing, frightened Iranians have handed over their savings in a French back alley after crossing the whole of Europe in search of freedom and somewhere to belong that isn't home.

I thought about how, despite my daily summertime immersions, the Channel is and will always be utterly indifferent to me. As soon as I leave it, all trace of my presence is gone. Even my footsteps in the shingle are washed away by the tide. I shudder at the thought that it would take only a few strokes too many towards the horizon at the wrong time on the wrong tide in the wrong weather conditions, and I'd become another footnote in the Channel's long and harrowing history of disappearances and drownings.

Although the sea where I swim is about as gentle a stretch as you'll find, on mornings when the wind's from the east, even here with the protection of the Goodwins, the Channel can get lumpy and I find myself taking facefuls and mouthfuls of sea water as wave after wave comes at me and I rise and fall with them, gasping to keep my face above the surface, conscious of the power of the sea even a few short yards from the shore. The feelings of immersion and space that I relish turn into a watery claustrophobia as the waves mass around me. When I turn towards the beach the understated power of the sea carries me forwards in a way that makes swimming useless and reminds me of my own utter insignificance. Yet on other mornings the sea can be almost perfectly still, the surface as smooth as silk and the colour of mercury, the sky covered by a thin layer of cirrus and the morning haze hiding the horizon to make sea and sky indistinguishable.

For my first couple of swimming years, the end of the summer when the mornings grew darker and the temperature of both air and sea began to fall spelt the end of my season. Shorts, wearable towels and sports cloaks hunkered down for the winter in the bottom of the wardrobe. By my third summer as a sea swimmer, however, something had changed. As autumn arrived with its fogs and rains I found myself not wanting to give up the Channel for the winter. I wanted to keep going, to keep swimming, even on cold, dark mornings when the combination of a many-togged duvet and central heating made heading out into the sea clad in just a pair of shorts seem like the most ridiculous thing in the world.

This would be my Channel winter. I'd not just keep swimming

through it, I'd spend it exploring too, the places, the people, seeking out its stories and trying to find what makes this stretch of water so very different from any other. To start doing that, I needed to find out where it came from.

4

The Channel Story

The English Channel is a whelp, a wean, a whippersnapper, a sea so young it barely registers as a blip at the end of a timeline of the history of the planet. If we think of the planet in terms of your life, the moment you were born representing the formation of planet Earth, the English Channel appeared roughly when you started reading this sentence. It's quite a long sentence, but still.

Half a million years ago there were rolling hills where the Channel is today. They made up part of the Weald–Artois Anticline, which might sound like the name of one of Jacob Rees-Mogg's children but is actually a ridge formed by an outlying reverberation from the ancient collision of the African and Eurasian continents called the Alpine Orogeny, an event that gave us the mountain ranges of Europe and the Caucasus. The part of the ridge in our neck of the woods ran roughly north-west to south-east and can still be discerned from the broad line of the South Downs and the corresponding landscape that stretches away inland from Calais. What's now the Strait of Dover was once the highest point of the Weald–Artois Anticline. There were humans on our continental protuberance around 800,000 years ago if the stone tools and imprints of footprints found at Happisburgh on the Norfolk coast in the early 2010s are anything to go by. The footprints, which were exposed by coastal erosion and washed away two weeks later, were thought to have been made by *Homo antecessor*, a people of limited intelligence who lacked a strong chin and whose traces you can still see today in our aristocracy.

Homo antecessor might recognise aspects of Happisburgh today, but the big difference they'd notice is that you can no longer enjoy a long coastal ramble to Rotterdam. At Happisburgh you'd have also been at the mouth of the Thames, which along with the Scheldt, Meuse and Rhine was happily emptying fresh water and silt into the sea. Which was all fine, and people in Happisburgh spent a good 300,000 years chipping away with their stone tools and leaving footprints around the place and were happy doing it, until one day around half a million years ago a stone-chipper stopped chipping, looked up at a footprint-maker and asked whether it was just him or was it getting a bit chilly? Global temperatures were in the process of dropping by around two degrees, causing the polar ice cap to make its way a considerable distance south, as far as Swindon, at which point the stone-chippers and footprint folk vanished.

As the sea was gradually pulled away, absorbed into the spreading ice cap, a glacial lake formed in the southern half of what's now the North Sea between the ice sheet and the Weald–Artois Anticline. With the rivers still emptying into the lake, this natural basin kept on filling until at some point around 450,000 years ago there was a major inundation, water from the lake brimming over the top of the Anticline and crashing down into the sea on the other side. A huge gash in the seabed was created thirty feet deep and nearly twenty miles wide. Geologists who have studied the impact estimate the equivalent of the entire modern North Sea washed over in the space of just two weeks – far too much water to be accounted for simply by the geological equivalent of an overflowing sink, but nobody has yet been able to explain where all that water came from.

Over the next 275,000 years the water level dropped again

and we were left with a band of slightly marshy but fertile land between Britain and France, threaded with rivers and streams, its size fluctuating with changes in the global temperatures, tectonic shifts deep within the still-settling planet and the compacting of sediment, but even at its narrowest it remained an estimated 300 feet high and twenty miles wide in a stretch roughly between Dover and Calais. The Neanderthals made great use of it, proving to be better hunters than the Happisburgh people had been.

Gradual adjustments in the global climate continued as the Earth got used to being a planet, and as it warmed and cooled sea levels continued to rise and fall accordingly. All the while that stubborn stretch of land remained persistently above the waves. It was gradually sinking, though. In fact the floor of the North Sea is still going down at the breakneck rate of about ten metres every 100,000 years: if you go for a paddle at Skegness and stand really still you can probably feel it. During warmer eras of high sea levels the stretch became too marshy to cross; people could only traverse it during the cold periods. Eventually, possibly for this very reason, the transient peoples of ancient times gave up on the region altogether and hence there was nobody – not even *Homo sapiens*, who had started throwing his weight around other parts of the continent at that stage – to see the final formation of the English Channel. Which is a shame, as it must have been quite something.

It was French engineers surveying the seabed ahead of the Channel Tunnel works during the 1970s who first noticed a deep, sand-filled depression beneath the Strait of Dover that they called the Fosse Dangeard, or the Dangeard Pit. Belgian geologists investigated further and found it was just one of a series of depressions stretching across almost the entire seabed between Britain and France, vast scoops out of the rock that were as deep as five hundred feet beneath the surface of the

Channel. They were all filled with sand and silt, all in a line corresponding with the stubborn Weald–Artois Anticline and deeper at their northern ends, and all bearing exactly the same characteristics as the plunge pool at the bottom of a waterfall. These were in fact 'some of the biggest plunge pools we've ever seen on Earth', according to one of the Belgians who examined them. They proved to be much more recent than the inundation of 450,000 years ago: evidence of that was still visible, showing that these gouges in the seabed had to be more recent. Much more recent, as it turned out.

By around 180,000 years ago an ice sheet had again crept as far south as Swindon, blocking the movement of water around the north of our islands. As the rivers continued to empty into the lake off our east coast between the bottom of the ice sheet and the northern coast of our Anglo-French land bridge, the less substantial nature of the ground – mainly compacted sediment rather than rock – ensured that this time when a breach came it was much greater, the rush of water first creating falls half as high again as Victoria Falls and several times wider, stretching the valley between Britain and France to something approaching the modern dimensions of the Strait of Dover.

It was still possible to walk between Britain and France for most of the last 125,000 years, however, thanks to the continued fluctuations in global temperatures. The Channel and the southern North Sea are among the shallowest in the world, which meant that cold periods would see the sea shrinking away into the ice sheet again, restoring the continent to its former complete glory. This time, instead of emptying into a giant North Sea pool, the Thames, the Rhine, the Scheldt and the Meuse fed into a river whose course followed the crevice formed by the original ancient inundation. The Neanderthals liked the look of this and returned around 60,000 years ago. Twenty thousand years after that, *Homo sapiens* arrived after a particularly cold spell had left uninterrupted land stretching between Britain and as far north as modern Denmark, an expanse with vegetation and rivers that was given the name of Doggerland.

We have a house-proud lobster to thank for much of what we know about the people of the Channel around this time. Fishing boats had been picking up flint tools and animal bones from the Pleistocene era in their nets since the nineteenth century, but in 1999 divers about forty feet down on the seabed close to the cliffs at Bouldnor on the Isle of Wight noticed this particular crustacean pushing items out of its burrow that didn't look like regular bits of rock. On further examination the items dumped on the doorstep by the lobster turned out to be stone tools from the Mesolithic era dating as far back as 15,000 years ago. Archaeologists already knew there had been a forest under the water there, but the tools discarded by the lobster showed that the humans in the region before the inundation were at least 2,000 years more advanced than previously thought. They even found traces of wheat dating from the period, a crop that wouldn't be grown in Britain for another couple of millennia.

Inevitably temperatures began to rise again, raising global sea levels by as much as 300 feet as glaciers melted and the water gushed back into the oceans, a process that continued until as recently as 5,000 years ago when the Earth began to look largely as it does today. It was during this most recent transformation that the English Channel we all know was formed. Approximately 8,000 years ago there was a huge undersea earthquake north of Doggerland close to the coast of Norway, a shifting of tectonic plates so catastrophic that massive landslides were triggered that dumped something like 3,000 square kilometres of rock into the sea. The surge that resulted turned into a wall of water five metres high, thundering across Doggerland towards the Strait of Dover where it shunted the last of the rock and sediment between Dover and Calais all the way through into the Atlantic.

It wasn't called the English Channel then, of course, as there wasn't an England. Indeed it would take until the early nineteenth century for there to be any sense or unanimity in the English naming of our nation-defining waterway. As far back as the sixteenth century it was appearing on Dutch maritime charts as the *Engelse Kanaal*, but even as late as the 1820s you were as likely to read in the newspapers about the British Channel as you were about the English. The Greek mathematician and geologer Ptolemy had called it the equivalent of the British Sea on his maps in the second century AD, while the seventh-century Andalusian philosopher Isidore of Seville referred to an *Oceanus Gallicus* in his influential *Etymologies*. In the Middle Ages the Channel appeared as the *Mare Britannicum*, *Mare Gallicum* and even 'the Sea of the Coast of Gaul' on charts produced in various parts of Europe.

According to the *Oxford English Dictionary* the first written instance of the name 'Channel' in reference to the sea between Britain and France occurs in the works of Shakespeare. In *Henry VI Part 2*, believed to have been written sometime between 1591 and 1593, William de la Pole, Earl of Suffolk, has been sent into exile for his role in the murder of the Duke of Gloucester, but the ship taking him to Calais is captured off the Kent coast by the vessel *Nicholas of the Tower* on which Suffolk is put on trial by the captain for his treachery.

On receiving the death sentence an understandably arsey Suffolk rails angrily at the captain and his crew. 'Thy words move rage and not remorse in me,' he fumes. 'I go of message from the queen to France; I charge thee waft me safely cross the Channel.'

Perhaps more significant than the naming of the Channel is the way in which it's been regarded through history, particularly from our side of it. Perhaps surprisingly, the notion of the Channel as a moat cutting us off from the foreigners across the water is a relatively recent development, dating back roughly as far as the eighteenth century. While the seas around us to the north, east and west played a protective role, the Channel was seen more as a facilitator of communication and connection. St Gildas, a sixth-century British monk best known for writing a history of Britain before the Saxons arrived (the cheerfully titled *On the Ruin of Britain*), noted how the seas around us created a significant and protective barrier except for the one on our southern shore.

'It is protected by the wide, and if I may so say, impassable circle of the sea on all sides,' he wrote of Britain, 'with the exception of the straits on the south coast where ships sail to Belgic Gaul.' Geoffrey of Monmouth, writing nearly half a millennium later, echoed this view, describing how Britain's 'straits to the south' would 'allow one to sail to Gaul'. And while there were undoubtedly Channel crossings in very ancient times – Dover's impressive museum has an extraordinarily well-preserved Bronze Age boat on display whose dimensions suggest it was capable of reaching the land its builder could see across the strait – the first person we know for sure sailed in the Channel was Pytheas.

A geographer from the Greek colony of Masillia, modern Marseille, Pytheas made a voyage around north-western Europe in around 325 BC and wrote an account of his adventures called *On the Ocean*. The document itself is long lost, so that we only know about it through the writings of other people, which is a shame as it sounds great. *On the Ocean* was notable for being far more than a periplus, the kind of basic navigational record that prevailed at the time, containing instead a range of information from astrological observations to personal anecdotes – it was one of the earliest travelogues ever written, in fact. Pytheas's vessel passed through the Straits of Gibraltar, followed the coast of Western Europe and may even have circumnavigated the entire island of Britain. In the fragments quoted in other works, Pytheas mentions a place on the south coast he called Belerion where the locals mined tin that was traded with Gaul, suggesting he was in Cornwall, as well as referring to Kantion, Kent, and Orkas, which seems likely to have been Orkney.

Pytheas is quoted extensively in Strabo's *Geographica*, written up to 300 years after the voyage, but he is included there only because the author is determined to expose Pytheas as a fibber on a massive scale. He was 'the worst possible liar' according to Strabo, who dismissed his account as a litany of fabrications. A couple of hundred years earlier Polybius had penned a similar panning of *On the Ocean*, but despite these comprehensive literary eggings Pytheas's reputation remains largely intact. Pliny's *Natural History* cites him as an unimpeachable source, as does the historian Timaeus, both writers producing histories that drew on and praised Pytheas, significant particularly in the case of Timaeus, who was writing while the older writer was still alive.

Certainly, Pytheas seems to have accurately estimated the circumference of Britain as around 4,000 miles, and his stab at the distance between northern Britain and his hometown Marseille wasn't far off either. It was Pytheas who launched the theory of Thule, the mysterious land to the north that could have been Iceland or Norway, and he was the first recorded person to see the midnight sun and to notice that tides are affected by the moon. He must have been a hell of a seaman and a remarkable navigator: the voyage would have been challenging enough today with the advantage of weather forecasts, charts and GPS, especially in the flat-bottomed vessel in which he most probably sailed, but Pytheas literally had no idea where he was going or what he might find when he got there. Perhaps most impressive of all, despite the reservations of Strabo and Pybius, he doesn't appear to have just made stuff up.

No one knows for sure what kind of vessel he used for his epic voyage. As he was most likely a merchant looking for tin and amber it could have been a flat-bottomed *holkas*, a cargo boat, sleeker than the famous triremes but probably not much fun when it was being tossed around on a heaving Channel. What we know of his writing still seems to add up today, more than two millennia on, and his account of Britain – which he described as densely populated, extremely cold and ruled by many kings and nobles, which sounds pretty much on the money even now – was the only recorded source describing the island until the Romans showed up.

Before Julius Caesar arrived on the Channel shore with a notion to conquer Britain there had been much toing and froing across the Channel, not least any number of Gallic refugees escaping the Roman advance. The south-Brittany-based Veneti tribe had traditionally held effective control of the western Channel, exercising a maritime dominance with their sturdily built ships with leather sails that allowed them to charge tolls from those seeking to cross. The Romans eventually overcame the Veneti at the naval Battle of Morbihan in 56 BC, the decisive moment coming when the wind suddenly dropped completely, allowing the legions to board the larger Veneti vessels with relative ease. 'Nothing could have been more fortunate,' Caesar noted.

A number of Gallic itinerants became chieftains in Britain well placed to offer assistance to their brethren across the water, something that made Caesar determined to bring the Britons under control and secure the northern borders of his empire. He arrived at Boulogne after Morbihan and began to assemble a large force of legionaries close to the headland at Cap Gris-Nez, while ordering his victorious fleet of ships up the Channel from Brittany to join them. The first significant invasion of Britain from the south was about to begin.

A few yards from where I live there is a small monument in the style of a stone Roman coin commemorating the exact spot where in 55 BC Caesar himself landed in Britain for the first time, a little bit by accident. His plan had been for two legions to embark from Boulogne in eighty ships with a further eighteen bringing the cavalry from a point somewhere west of Cap Gris-Nez. The legions embarked as planned, but by the time the cavalry had loaded their ships they'd missed the tide and were forced to turn back. Caesar's fleet, meanwhile, having sailed through the night, arrived off the coast of Dover as the sun rose over low tide on 26 August, to be greeted by locals gathered on the cliffs making it clear they weren't overjoyed to see him. Caesar waited for the tide to rise, then sailed along the coast until he spotted another possible landing place and settled on Deal. The problem was there were yet more angry-looking people waiting for them on the shore and, with the fleet grounding some way short of the waterline, the legions had to jump into the sea with all their equipment and wade ashore before even thinking about fighting. The locals charged up and down the shoreline roaring and shouting – something you don't get in refined Deal today, let me tell you – and an understandably reluctant standard-bearer needed a few moments before he dropped into the water and led the way ashore.

After some heavy fighting the Romans managed to gain the beach, but without cavalry support they couldn't risk advancing further inland so made camp instead. The cavalry ships finally sailed into view only for the weather to close in, with fierce northerly winds eventually forcing them all the way back to Gaul. The storm lasted for days and coincided with a spring tide, the rising seas causing many of the Roman ships to drag their anchors and end up wrecked on the beach. After another skirmish with the locals Caesar decided he had seen enough and went back across the Channel.

He returned the following year, this time determined to make sure, with 600 ships carrying five legions. He made it further inland, as far as the north side of the Thames, but when the weather turned again, wrecking more ships at anchor, he decided to return to Gaul while he could rather than risk spending a long, hard winter in Britain. So keen was he to go home that he left no forts or garrisons, elicited just a few vague promises from local chieftains to behave themselves and took a few cursory hostages back with him. There would be diplomacy and trade in the meantime, but it would be nearly ninety years before the Romans under Claudius set about making Britain a fully fledged outpost of the Roman Empire.

For the first time in recorded history the English Channel had proved a formidable barrier to an invading force. The Romans were used to the gentler seas of the Mediterranean rather than the stroppy waters of the Channel in autumn, with its big tides, pulls and eddies, not to mention weather conditions that could change in an instant. Caesar had been seduced at Cap Gris-Nez by just how close Britain looked, how harmless the body of water in between seemed. He could even see settlements on top of the cliffs, and the water in the summer sunshine would have looked turquoise and calm enough to have him pining for the waters of Italy again. Yet on both occasions when he brought his fleet over, the weather caused decisive chaos, rain-strafed gales driving ships into each other and onto the shore. The Channel had pulled off its first recorded deception, the old harpie, one that would hoodwink potential invaders for centuries to come.

A few of them would be successful, the old notion that we've not been invaded since 1066 being well wide of the mark, as there have been more than seventy invasions of varying sizes and degrees of success. In 1688, for example, William of

Orange sailed along the Channel and was virtually unopposed when he landed an army in Devon – after which, a skirmish at Reading aside, he practically sauntered into Windsor nodding at passers-by, en route to assume the English crown once James II had fled to France. Yet we don't call that an invasion: we call it the Glorious Revolution.

In 1217 we'd also happily waved in an invading force from the Channel, when King John decided that having been forced to sign Magna Carta in 1215 he no longer felt obliged to abide by it, triggering a war between the king and the rebel barons who'd strong-armed him into applying his seal to the charter at Runnymede. When John looked like he was winning, the barons appealed across the Channel for help and petitioned Louis, the heir apparent to the French throne, for his assistance. Louis seemed only half-interested at first, sending some knights over to help defend London, but the more he thought about it the better he liked the idea of putting together an invasion force, despite the disapproval of his father Philip II, not to mention the pope.

Hence in May 1216 Louis, in defiance of just about everyone worth listening to, set out across the Channel with a fleet of ships and disembarked an army somewhere on the east coast of Kent. Hearing of this, John decided to leave London with his court and hole up in Winchester, allowing Louis to march his army to London unopposed, where he was welcomed by the barons, taken to St Paul's Cathedral and proclaimed king in lieu of a proper coronation shindig.

In October that year King John died of dysentery at Newark at the age of forty-nine, instantly making Louis seem to the barons less of a solid-gold candidate for coronation than John's nine-year-old-son Henry, waiting quietly in the wings. Henry had two advantages in the eyes of the barons: he represented an uninterrupted royal succession and he wasn't John. Louis meanwhile had been passing the time besieging the odd castle around the south of England and thinking of bands he could get to play at his coronation. Then he received the news that not only was he no longer the red-hot favourite for the throne, but his former champions were now heading his way armed to the teeth. He hot-footed it to the coast where, having come through a scrimmage at Lewes, he was plucked from a tricky battle situation at Winchelsea by the timely arrival of a French fleet that hauled him back across the Channel.

This slight setback – practically the entire country not wanting him to be king any more, a point of view echoed by his own family and the Church – didn't dissuade Louis, who was determined to go back to London and have another crack at the crown. Having assembled an army he crossed the Channel and set about conducting a siege of Dover Castle, something he'd tried on his previous sojourn without success. Dover folk are hardy people and their castle is a particularly hardy structure, and the siege drew so much of Louis's resources that the handful of castles his troops managed to take elsewhere fell one by one.

In the meantime a French fleet set off from Calais on 24 August led by the *Great Ship of Bayonne* under the command of one of the great historic Channel characters, Eustace the Monk.

'No one would believe the marvels he accomplished, nor those that happened to him many times,' was a near-contemporary verdict on the man born Eustace Busket in 1170. His father Baudoin Busket not only had a name that sounded like a minor character in a P.G. Wodehouse novel, he was also a lord of Boulogne. According to apparently reliable sources, at a young age Eustace travelled to Spain to study black magic at Toledo, where the dark arts were so openly practised that Head of Black Magic was a respected and popular post sought after by local academics at the town's seminary of magic. Eustace, it was reported, spent six months there studying necromancy, the art of communicating with the dead. In a bit of a career shift he returned home to become a Benedictine monk at an abbey outside Calais, but on earning himself a reputation there for gambling and inventively foul language, he left after his father was murdered, vowing to track down his killer.

By the early years of the thirteenth century Eustace had become right-hand man to the Count of Boulogne, Renaud de Dammartin. The two fell out in 1204 to the extent that Renaud had Eustace declared an outlaw and confiscated his lands. Some accounts say that Hainfrois de Hersinghen, the man apparently behind the murder of Baudoin, turned Renaud against Eustace; either way, the now ex-monk disappeared into the Forest of the Boulonnais and commenced a Robin Hood-style guerrilla campaign against his former employer.

Much of this period is recorded in the *Romance of Eustace*, a late-thirteenth-century epic that details the outlaw's persistent humiliations of Renaud, frequently stealing his horse pretty much from under him while employing various disguises including that of leper, mackerel seller, potter and – get this – prostitute. It wasn't all laughs, though: so genuine was Eustace's sense of grievance that when his nemesis threw his support behind the King of France, Eustace, who had also been indulging in a little light Channel piracy on the side, promptly sailed to England and offered his services to King John.

Impressed by his piratical credentials, in 1205 John gave Eustace thirty ships to employ, as he saw fit, against Philip. Eustace proved an effective mercenary, establishing himself in a castle on Guernsey and successfully invading Sark, with many accounts claiming that John subsequently awarded the Channel Islands in their entirety to Eustace, who already had a well-appointed house in London as well as his most frequently used residence in Winchelsea. In 1212, however, Renaud turned the whole situation on its head by suddenly allying himself with

King John, immediately prompting Eustace to switch sides the other way. When John launched an assault on the Channel Islands he responded by setting about Folkestone. When the barons began their post-Magna Carta agitation against John and invited Louis to England, it was Eustace at the tiller of the boat that sailed him across. Inevitably it was also Eustace who extricated Louis from the tricky situation at Winchelsea, and Eustace who, to aid Louis's original campaign, organised and oversaw the landing of military equipment and supplies on the French coast.

On 24 August 1217 Eustace sailed out from Calais at the head of his fleet on the *Great Ship of Bayonne* to bring aid to Louis once again. His own vessel was so heavily laden with horses and supplies it sat low enough in the water for the sea to wash over the decks, not an ideal situation. When approaching Dover he saw the English fleet of Hubert de Burgh heading out to meet him. A fearsome scrap followed, one of the most intense sea battles ever fought in the Channel and one that carried the ships up along the Kent coast towards the Isle of Thanet. Eustace's heavily laden craft gave as good as it got until it found itself caught at close quarters with an English vessel. Despite being heavily outnumbered Eustace's crew put up an extraordinary resistance in which the skipper inevitably took a leading role.

'Eustace himself crushed many with the oar he wielded, breaking arms and legs with every swing,' reads one account. 'This one he killed, another one he threw overboard. This one he knocks down, another he tramples under foot, and a third one has his wind-pipe crushed.'

It was only when the English began rather unsportingly lobbing barrels of lime onto the French ships, sending up blinding clouds of choking dust, that the fight began to swing their way. Inevitably, the lime-covered French were overpowered and Eustace was found below deck hiding in the bilges. He offered to serve the English king, along with 10,000 marks in cash in exchange for his life, but it was all in vain.

'There was one there named Stephen of Winchelsea,' reads one epic record of the battle, 'who recalled to Eustace the hardships he had caused them on land and sea and who gave him the choice of having his head cut off either on the trebuchet or on the side of the ship. Then he cut off his head.'

Eustace's severed head was taken back to England, stuck on the end of a lance and dispatched to Canterbury, from where it departed on a nationwide tour as a deterrent to anyone thinking of taking on the new regime of Henry III. It was a grisly conclusion to a remarkable life – indeed, one of the great Channel lives.

Meanwhile, cut off from his supply line across the Channel by the death of Eustace and the defeat of his fleet, Louis had no choice but to give up. By relinquishing his claim to the crown of England he was allowed to return to France, with the barons who had lobbied for his invasion of England handed the bill for sending the French prince and his considerable entourage back across the Channel.

If the death of Eustace the Monk marked the first great battle in the Channel, the first major disaster had taken place a century earlier, a tragedy whose ramifications were still being felt as Eustace battled the English.

The *Blanche Nef*, the White Ship, was arguably the greatest vessel of the age as she sat in the harbour at Barfleur in November 1120. With her carved prow, single mast and huge sail, not to mention the pristine white paintwork that gave her her name, the 120-foot vessel was considered the most beautiful afloat. Captained by Thomas FitzStephen, whose father had been in charge of the *Mora*, the Norman flagship that brought William the Conqueror from Normandy to Pevensey at the head of a fleet of 700 ships in the late summer of 1066, the recently renovated vessel was under the command of the greatest seaman of the age. With her crew of fifty, most of them oarsmen, the *Blanche Nef* was capable of accommodating several hundred passengers, plus a decent amount of cargo, so when FitzStephen offered his and the ship's services to Henry I of England to transport his sons William the Aetheling, the legitimate heir to the throne, the illegitimate Richard and his illegitimate daughter Matilda, Henry was happy to grant the request.

'I entrust to you, Thomas FitzStephen, my sons William and Richard whom I love as my own life,' he told the veteran skipper. The monarch had already arranged his own transport and set sail from Barfleur in the late afternoon of 25 November. His son was more than happy to stay on a little longer: with the King departed he was suddenly presented with some bonus drinking time, free of his father's disapproving frowns.

William represented a great deal more than simply the next King of England, as if that wasn't enough: he was the grandson of William the Conqueror, while his mother Matilda of Scotland was a great-niece of Edward the Confessor. William represented a bright future for Anglo-Norman cooperation. Although he'd not long turned seventeen he had just been made Duke of Normandy and had married Matilda, the daughter of Count Fulk of Anjou, another politically expedient match that boded well for a peaceful future. Henry was away often and would leave his wife as regent, but following her death in 1118 William had effectively been king designate. At Barfleur that night he'd just accompanied Henry on a mission to negotiate peace with the bumptious Louis VI of France and was already establishing himself as an important political figure on the European stage.

Unfortunately he was also a colossal dick. One Norman chronicler referred to his 'immoderate arrogance' while in his *Historia Anglorum* Henry of Huntingdon refers witheringly to how 'pampered' the prince was. An estimated 300 people were on board besides the crew that evening, and all of them seemed determined to get right on it in a big way. Perhaps thinking they wouldn't sail that night, many of the crew also got thoroughly stuck into the good stuff and before long everyone was absolutely blootered. When a priest arrived at the quayside intending to bless the ship ahead of her voyage he was sent away, leaving wide-eyed at the bacchanalian scenes he had witnessed on board.

Then the evening took its fatal turn, as William noted the strong southerly wind and decided they should sail immediately, so as to beat his father home, for a laugh. It was a really, really terrible idea, but FitzStephen had to balance his maritime instincts against the fact that the pissed-up entitled brat in front of him was the heir to the throne. He ordered the crew to make the vessel ready to sail.

It was around midnight on a clear night when the *Blanche Nef* weighed anchor, unfurled her sail and nosed out of the harbour into the Channel, dark beneath a new moon and with the whoops, hollers, songs and belches of the nation's finest carrying across the water. No sooner had she cleared the haven of the harbour than the ship, already picking up considerable speed, smashed into a notorious rock called the Quilleboeuf, ripping a fatal hole in the vessel's timbers. Immediately there was panic. In the dark, lamps were knocked over or were snuffed out by the water as the ship listed and passengers tumbled into the sea. Within minutes she was on her side and everyone was in the water. FitzStephen managed to get William into a boat with a couple of crew members and dispatched them shoreward, until the prince heard his sister's screams in the water and insisted on going back to look for her. When the thrashing mass of desperate people surrounding the boat tried to clamber aboard they succeeded only in capsizing her, sending the future king head first into black water that concealed fatal dangers.

'Instead of wearing a crown of gold,' lamented Henry of Huntingdon, 'his head was broken open by the rocks in the sea.'

In the drunken panic and chaos of that Barfleur night everybody on board, all 350-odd crew and passengers, nobleman and commoner alike, drowned except one, a butcher from Rouen called Berout who'd gone on board looking for payment for some meat he'd supplied and found the ship sailing with him on it. He managed to cling onto a piece of wreckage all night before washing up exhausted on the shore at dawn, the only witness, the only person to pass on the tale of the Channel's first great disaster.

Only a handful of bodies were recovered, William's not among them. His wife had sailed earlier with Henry and when the news of the disaster reached England both she and the king collapsed onto the floor. After a few months at Henry's court she returned to Anjou and took holy orders, spending the rest of her days in a convent. Henry, left without an heir, married again but produced no more legitimate children, triggering a crisis of succession that rocked European politics for the next three decades, a period known as 'the Anarchy' whose effects were felt well into the next century.

The English Channel claimed the nobles, the princes, the courtiers, the advisers – nearly the entire English court and a hefty chunk of the establishment – in a few desperate, thrashing, panicked minutes in a dark sea on a chilly night off the coast of Normandy. And with the White Ship tragedy began the Channel's litany of loss: the Spanish Armada, the Normandy landings, the *Herald of Free Enterprise*, the Penlee lifeboat, Emiliano Sala.

It was time for me to cross that dangerous stretch for the first time since I became a Channel person. I'd start with the piece of land I could see in silhouette across the water in the morning sunshine, long and low like a whale, but which in the evening when the sunlight came from the west became bright-white cliffs with green tops, exactly like our own.

5

Calais

I was the only person on the road to Calais. There was a certain joyfulness about the vehicles that disembarked from the *Spirit of Britain*, like cattle let out of their sheds after the winter. The brief purgatory was over and spirits rose in tandem with each move up through the gears. Whatever journeys people were on, whether heading east for the Low Countries and Germany or south to the rest of France, they were properly under way now.

My journey was the shortest of them all. As the snaking traffic eased straight over the roundabout at the port exit and onto the N216, I practically doubled back, taking the last exit and nosing along Avenue du Commandant Cousteau heading for Calais's elaborate neo-Flemish tower of the Town Hall beyond, whose clock face can be read on a clear day with a pair of binoculars from the cliffs above Dover. After kerclunking over the metal bridge on the Calais and St Omer Canal and parking on the quay by the marina, I was checking in to my hotel barely ten minutes after disembarking from the ferry.

I might have been the only one doing it this time, but as a British visitor to Calais I was continuing a tradition dating back hundreds of years. Most of us travelling to France now pass by the town almost without thinking, especially since the Age of the Hypermarket drew to a close (in 1999 when Calais was the cheap booze and fags capital of Europe one-third of the people in the town at any one time were English), but in times past

Calais was either a destination in itself or somewhere to recover from the crossing before continuing into the rest of Europe. Not to mention the fact that for a good couple of hundred years or so in the late Middle Ages Calais was part of Britain.

I'd brought a guidebook with me, *An Englishman's Guide to Calais* by James Albany. The reviews looked good. 'A very useful and amusing directory for the tourist,' said one, and I couldn't reasonably ask for much more than that. The only possible drawback was that, like the review, *An Englishman's Guide to Calais* was published in 1829 so there was a fair chance it wouldn't be entirely on the money with prices and opening times. Nevertheless, once I'd reached my room I parked myself supine on the bed and immersed myself in James Albany's Calais.

He really, *really* liked the place. As early as the first page he was lamenting Britain's loss of it, a loss that had taken place almost 300 years before but was still raw enough for James to rail against the 'shameful neglect which led to its recapture by the Duke of Guise'. He thinks he knows whose fault it was, too. 'Well might Queen Mary exclaim that when she died Calais would be engraven on her heart,' he tut-tutted.

The loss of Calais in January 1558 made a terrific dent in English self-esteem, one whose reverberations were still clearly being felt in the early nineteenth century, if Albany's sulk was anything to go by. Since Edward III had taken Calais in 1347 the fortified town had been a strategic blessing, aiding the export of English wool to Flanders and beyond as well as being a handy base for keeping an eye on Europe and its wars from a fortified vantage point, without necessarily having to join in. So English was Calais back then that its streets bore English names, and from the reign of Henry VIII it had even returned MPs to the London parliament.

Mary Tudor was on the throne in 1557 when in December

her husband Philip of Spain, at the time engaged in the kind of squabbly war with the French that the Channel generally kept us out of, tipped her the wink that the French were planning an attack on the town. Traditionally, Calais had been left alone by the various warring factions as they preferred the English being there to whoever it was they were fighting that week, so this news took Mary rather by surprise. She ordered the Earl of Rutland across the Channel with as many soldiers as it would take to back up the existing garrison under Thomas Wentworth, who for years had been pointing out the vulnerability of Calais's obsolete defences, a mitigation that wouldn't keep him out of the Tower of London when the town fell. A flu epidemic had rendered most of England's fighting men practically useless, meaning that by the time Rutland set out across the Channel on 2 January 1558 with boatloads of pale soldiers parping into their handkerchiefs, the Duke of Guise had taken advantage of the newly frozen marshy coastal land around the walled town to skitter over it unopposed and cut Calais off from its own harbour. From there he launched an intensive bombardment that saw Wentworth surrender the keys after the briefest of sieges. It was around this point that Rutland finally showed up, asking if anyone had any Lemsips.

Among the practicalities advised by Albany before one departed for Calais was the procurement of a passport. These could be obtained by calling on the French Ambassador in London at the embassy on Portland Place between 1 p.m. and 3 p.m. weekdays and leaving your name and address. Your passport would be ready for you to collect the next day and wouldn't cost you a penny. Indeed, so relaxed was the issuing of passports in the late 1820s you could even send a friend along in your place 'if personal attendance should be inconvenient'.

The English were never huge fans of the French insistence on a passport to get around, incidentally, and regarded it as a bit of pointless bureaucracy. In 1849 when Louis Napoleon had become president of France following the 1848 Revolution, an editorial in *The Times* called on him to scrap passports altogether on the grounds that it would 'add indefinitely to the comfort and convenience of all foreigners'.

Once in possession of their passport the traveller to Calais had two options: a stage coach via Dover or, like Mary Capper, the steam packets that departed from the Thames beneath the Tower of London. The latter journey was estimated at 'twelve hours in fine weather, but travellers should allow for anything over eighteen in stormy conditions'. Government-run English packet-boats plied the Dover–Calais route for a fare of half a guinea, while the French mail boat sailing three times a week charged five shillings – although, Albany advises, 'they will often take less'.

These practicalities dispensed with, *An Englishman's Guide to Calais* turns into a journal of Albany's own visit in August 1827. He sailed with a friend and the friend's daughter on a mid-morning voyage from Dover, which he spent under an awning on deck for the impressive two hours and twenty minutes it took the ship, the *Crusader*, to make it over to Calais. Fast it may have been, but it can't have been a completely smooth crossing as he appeared to arrive in a bit of a funk, describing the French hills that hove into view as they approached the port as 'sterile and ugly'.

Once he actually got to Calais itself his mood brightened immediately. Straight off the boat, he wandered into a bookshop with a reading room, of which he approved, and noted the proliferation of cafés all of which were equipped with billiard tables. He passed a pleasant hour or so in the Café Giuliani with his friend, where they played billiards watched by three 'mechanics' wearing white caps who, to Albany's delighted surprised, observed their game quietly and respectfully.

Calais

'How different would have been the case in an English tap room!' he gushed, adding, 'If I am to judge from the Calesian plebeians the lower classes in France, particularly the females, are infinitely more civilised than those of the same grade in England.'

The next morning, a bustling market day, he visited Calais's main square the Place d'Armes, which was full of women whom he proceeded to leer at: 'Never before have I seen such an assemblage of handsome plebeian females.'

He spent quite a while, it seems, gawping at them going about their market business in this square that he found almost as delightful as the ladies. 'It being a fine day everything appeared *en beau*,' he wrote, showing off the bit of French he'd already picked up. The buildings lining the square – which he described approvingly as timbered houses with large windows built of yellow brick with green trim – helped to lend the whole scene a 'gay appearance'.

While I could have passed the entire evening lying there reading Albany's guide to Calais, I closed the book, put it in my pocket and headed out to the Place d'Armes, the marketplace that was apparently attractive enough to distract Albany from the giddiness induced by the half-exposed wrist of a woman selling fish.

Throughout Calesian history the Place d'Armes has always been one of those multipurpose civic spaces, home to the market, a meeting place, somewhere proclamations might be read out and, especially during the British years, a popular venue for crowd-pulling executions. A string of English nobles were hung, drawn, quartered and beheaded on the Place d'Armes during the couple of centuries before the town was lost to their nation, mostly during the reign of Henry VIII and after for alleged plots of the popish variety. When you're a little fortified Protestant town backed up against the Channel in

47

an almost exclusively Catholic country you're probably going
to be a bit jumpy, but so many people fell under the behead-
first-ask-questions-later policy in the Place d'Armes that the
executioners of Calais became master practitioners of their art,
with a reputation that spread across Europe and the Channel.
For Anne Boleyn's dispatch to the next realm, they sent for the
city's most experienced neck-slicer at Henry VIII's request, in
order that his erstwhile wife's head be removed quickly and
with minimal distress, the old romantic.

The Place d'Armes today might not be soaked in the blood
and entrails of English nobles, but it's certainly lost the charm
that knocked James Albany through a loop. It's no more than a
vast expanse of paving surrounded on all sides by modern white
concrete boxes given a few windows to help them masquerade
as buildings. It's no longer a place to gather; it's a void, a place
to skirt around.

The focus and beating heart of the old Calais was almost
entirely flattened during the Second World War. It saw the
Place d'Armes replaced with a stony wasteland whose utter
absence of charm and character is almost impressive in itself,
even in the golden early evening sunlight of autumn when it
probably looks its best. For once, though, the transformation
of a square oozing history and character into a weary stand-
off between the assembled concrete monstrosities – their
ground floors a procession of bars and restaurants for which
the order seems to have gone out that in no way should they
attempt to blend in with each other – was unavoidable. Calais's strategic importance that had rendered it a vital piece of
English territory for 200 years almost guaranteed that it would
be razed during the war. First there was besiegement and
bombardment by the Germans in May 1940 in the days leading up to Operation Dynamo, then Canadian forces set about
German positions in and around the town in September 1944

on the grounds that you can't defend a place if it's no longer there.

The devastation left across northern France meant there was neither the time nor the money to think about restoring the town to its former glory, almost guaranteeing a soulless redevelopment so as to get the place back on its social and commercial feet as quickly as possible. That what should be the town's crowning glory is the worst example of all is a little bit heartbreaking, especially when you imagine the centuries of life, death, blood, fish guts and cabbage leaves trampled into its soil.

If there is any nod to the aesthetic, some remnant of the history of the Place d'Armes, it's in the south-west corner and takes the form of the sturdy, hunched shape of the Tour du Guet, the watchtower, standing some 130 feet high. There are some who contend it was originally constructed by Charlemagne in the early ninth century, although others date it to the early thirteenth and the fortifications carried out by Philippe Hurepel de Clermont. Either way, it's a welcome piece of heritage among the identikit postwar architecture of the rest of the square – not to mention its appropriate symbolism: Calais's position has traditionally been one of watchfulness against invaders from land and sea, and the tower continued to employ watchmen in times of both peace and war right up to the early twentieth century. Its survival is representative of the irrepressible spirit of Calais, its featureless surroundings emphasising the strength the tower radiates. It's a miracle the watchtower survived the last war, since just about everything else around it was flattened.

Indeed, the biggest threat to its existence wasn't an invading army at all, but a 1580 earthquake in the Strait of Dover that threatened to split it in two. So large was the quake that buildings in cities as far apart as London, Rouen and Lille suffered extensive damage, and houses and churches were destroyed in

Dover and Calais, but the Calais watchtower, possibly more than 700 years old even then, was still standing at the end of it. Since then it's been a bell tower, a lighthouse and even a telegraph station, from which news of the death of Napoleon Bonaparte was communicated to Britain in 1821.

In the shadow of the tower is a bronze statue of Charles de Gaulle strolling through Calais with his wife Yvonne. Unveiled in 2013, the sculpture is refreshingly unusual in its relaxed format, not to mention that it's a rare instance of a public figure being memorialised alongside their spouse. It's based on a 1959 photograph, and commemorates the fact that Yvonne de Gaulle was born and raised in the town. The positioning is significant too: the pair were married in the spring of 1921, a couple of hundred yards away at the cathedral-like Church of Notre-Dame, a building that also survived the war, and they're walking out of the square towards the Rue Royale, the closest Calais has to a high street and the erstwhile location of, arguably, its most famous establishment.

It's a warm evening as I walk down the Rue Royale away from the Place d'Armes, in search of a plaque: one that, had it been around in 1827, would have been faithfully transcribed by James Albany. The gaiety of the scene on the square had clearly exhausted his powers of description, as from then on he seems to spend most of his time noting down every sign he sees. Any kind of board or stone with writing on it anywhere in the town, and into the notebook it went. The names of shops and cafés, street names, warnings, direction signs: there was not a single piece of public lettering in Calais that went unnoticed. Ducking into a church at one point, he even makes a note of the signs telling people not to talk during the service.

This struck me as a bit of an odd affectation, until it occurred to me that, since his fairly obvious ogling of the local women probably wouldn't have gone unnoticed, he had been reduced to pretending that it wasn't the women he was interested in but the signs all around them. It was the Café du Grand Salon that fascinated him, he'd insist, not the rosy-cheeked Calesian plebeian standing beneath it – a claim he no doubt hoped would prevent his being bundled up an alley by an aggrieved husband or sweetheart and given a thorough going-over. A shave, the odd game of billiards and a visit to the theatre aside – Albany's visit to Calais seemed to consist entirely of eyeing up women and noting down signs.

I also wondered whether the signs were a necessary distraction because he was starting to feel, well, a bit *too* frisky around the ladies and needed something, anything, to take his mind off hanky and, indeed, panky. The signs were just the thing, and if there were none around when he felt the sap start to rise, then he'd have to make do with whatever he could lay his eyes on.

Take an incident he describes out on the sands with his travelling companion, when a group of twenty or so girls returning from a shrimp-gathering expedition came galloping towards them, all 'naked feet' and 'mirthful importunity', asking the fine English gentlemen if they might have a few spare *sous*. 'Luckily we had a good few of these coins which we threw among them for a scramble,' he said, a pretty rotten thing to do and one that may have aroused him to the extent that he needed something, urgently, to divert him. He wasn't exactly spoiled for choice, but he managed: 'They let out screams and ran around collecting the coins. I saw two bathing machines.'

When I found the plaque I was looking for, a small cement one about ten feet up on the wall of a pharmacy and dwarfed by the big green neon cross sticking out of the brickwork, I felt as I took out my notebook that it was in a way a kind of tribute to my guide – not least, knowing that he spent a large portion of his time on this exact spot.

'*À cet emplacement,*' I scribbled down, '*s'érigeait jadis l'Hôtel Dessin où, le 16 juillet 1831, SA MAJESTÉ LÉOPOLD 1er a séjourné avant de monter sur le Trône de Belgique.*'

Your French is unlikely to be worse than mine, but I can tell you the plaque relates that on that site stood the Hôtel Dessin where, on 16 July 1831, His Majesty King Leopold I stayed prior to taking the Belgian throne. Why the King of the Belgians gets the top billing here is a bit of a mystery – I can only presume the plaque was put up by some Belgians – because the Hôtel Dessin was for many years Calais's most famous establishment and the biggest draw in the town for the great and the good. Napoleon stayed here, as did Louis XVIII of France and George IV of England, not to mention Benjamin Franklin and a host of literary greats. Dessin's was a name known throughout Europe, as the travellers who stayed here always raved about the place and took their recommendations on their Grand Tours and business trips to every corner of Europe and beyond.

You know who also stayed there? James Albany. Of course he did. On his first night, before he'd had the chance to be captivated by the city and its women, he'd dined there with a large group of people including Monsieur Dessin himself, an experience he found excruciating having yet to throw off his English inhibitions. With most of the people around the table not knowing each other, and probably most of them being English too, the occasion was infused with that peppery social awkwardness we do so well, and at the conclusion of the meal everyone went their separate ways as fast as they could, Albany pausing only to record like the true Brit abroad that he found the mustard 'detestable, being strongly impregnated with garlic'.

Dessin's was a large hotel, big enough to surround two quadrangles with gardens that would later accommodate the building of a grand theatre, and it was already doing pretty well when the guest arrived who would change the fortunes of the place and assure its literary immortality.

In 1759 Laurence Sterne was a consumptive forty-five-year-old clergyman living in the village of Sutton-on-the-Forest a few miles north of York and trapped in a bored obscurity that was at least as unhappy as his marriage. Within a year he'd become one of the most famous names in the land, frequenting the best London clubs and coffee houses and counting the likes of Joshua Reynolds and David Garrick among a wide group of influential friends who actively sought his company and wit. Behind this startling transformation was the publication of *The Life and Opinions of Tristram Shandy, Gentleman*, Sterne's sprawling, bawdy beast of a novel that would eventually run to five volumes, published over several years.

Tristram Shandy was a sensation that catapulted the parochial vicar to the pinnacle of English literary society – and boy, did he take to it. 'I wrote not to be *fed*, but to be *famous*,' he told a friend, and he was overjoyed to be feted by Goethe as 'the most beautiful spirit ever active; anyone who reads him immediately feels free and beautiful'.

The boost to his clergyman's ego was matched by the boost to his clergyman's salary that came with his whopping royalties, riches that freed him to travel to London and beyond. He made his first visit to continental Europe in 1762, heading as far south as Marseille, then making a longer journey to Naples three years later. These travels inspired his second book, a volume that didn't garner quite as much attention as its mighty predecessor but which was just as revolutionary in style, format and subject matter. It also revolutionised the fortunes of a certain Calais hotel. In 1769 Sterne published *A Sentimental Journey*, a fictional realisation of the author's travels written in character as the parson Yorick, who also appeared in *Tristram Shandy*. It's a riotous work, effectively a collage of incidents and events and one destined to remain unfinished, as Sterne died in London in the spring of 1768 having completed two of an intended four volumes.

He'd stayed at Dessin's in 1765 and enjoyed himself very much, hence it was inevitable that on his sentimental journey Yorick would also hole up at the big place on the Rue Royale. Dessin had been playing off Sterne's fame even before the book was published, but with the hotel mentioned as early as the first page of *A Sentimental Journey* it became arguably the most famous in Europe, and stayed that way for the rest of the eighteenth century. Sterne's portrait by Joshua Reynolds hung over the fireplace, and the door of room 31 carried a wooden plaque declaring 'this is Sterne's room'. The literary parson would not, by a long chalk, be the only such luminary to rest his head under Dessin's roof – Thackeray (who wrote a short story about the place), Dickens and James Fenimore Cooper would be later guests, and when Sir Walter Scott stayed there in 1827 on his way to and from Paris collecting material for his *Life of Napoleon*, he asked specifically to have Sterne's room.

Dessin's reputation was not based entirely on Sterne's patronage, however. With eighty rooms, plus fifty beds available for servants, the hotel was renowned for its cleanliness, particularly of the bed linen. 'This house, though it has changed masters, is conducted as well as formerly, and there was nothing in it, which could have made the most determined lover of ease repent his having crossed the Channel,' was the verdict of an American officer of the Union Army called Ninian Pinkney in the early 1800s. In 1814 after the fall of the Empire, Louis XVIII spent his first night back on French soil at Dessin's, a landing so significant that a column commemorating the event overlooks the marina, with a brass footprint set into the base to mark that initial grounding of his royal size eight.

In the late 1780s, Christian VII of Denmark was the first

monarch to stay at Dessin's, and he was followed by numerous heads of state, including Leopold the King of the Belgians, whose plaque is the only physical trace now of Dessin's presence in the city. Napoleon Bonaparte was a guest as was his nephew Napoleon III, as well as Louis-Philippe I of France, and when George IV of Britain set off in 1821 to visit his territories in Germany he requested specifically a stay at Dessin's.

If you believed the English papers, George's arrival in Calais was quite something. 'Centuries may elapse before a spectacle of such interest recurs as that with which we were yesterday gratified,' panted a syndicated dispatch from France. 'The King of Great Britain landed in amity on the shores of France.'

He might have landed in amity, but at several points on the cusp of his arrival it looked like he might not land at all. The king had endured a pretty rotten crossing that was already longer than necessary because he had chosen to depart from Ramsgate, twenty-odd miles further north around the Kent coast from Dover, on account of the fact that he was still smarting over the rapturous welcome the traditional venue for cross-Channel departures had afforded his estranged wife Caroline of Brunswick the previous year. When he eventually reached Calais, the weather and the tide conspired to prevent the royal yacht entering the harbour, so a local boat was sent out into which the sea-groggy monarch clambered with some difficulty. The boat then grounded on a sandbank and by the time it lifted off again the king was thigh-deep in the water that was slowly filling the vessel. Finally he was able to reach the quayside, then proceeded on foot to the Place d'Armes and then on to Dessin's, keeping an eye out for a familiar face among the crowds lining the route, one whose presence betrayed the other side of the British in Calais – the refugees.

In February 1824 Lady Harriet Cavendish sat at her writing

table in Dessin's and described in a letter to a friend how Calais was 'peopled with English, slight sinners and heavy debtors, the needy and the greedy'. Seven years earlier she'd visited the city and found 'Lady Oxford and daughters living in a lodging in Calais, which seems to have become a sort-of purgatory for half-condemned souls'.

A combination of the absence of an extradition treaty between Britain and France and the much cheaper standard of living made France an attractive prospect for a skint Englishman, or as the nineteenth-century novelist George Reynolds put it, those 'obliged to leave their native land forever in consequence of the inhumanity of sheriffs' officers and policemen who would endanger their safeties under the paltry pretexts of debts or rogueries'. Philip Thicknesse, writing in 1776, was a bit more straightforward, labelling Calais 'the asylum of whores and rogues from England'.

Albany's guide didn't mention Calais's status as a haven for English refugees, but in listing its delights and conveniences it made a subtle allusion. 'If by all these appliances and means for enjoyment the traveller should still be visited by the demon ennui,' he concluded, 'he will always have the satisfaction of knowing that the matchless power of steam will secure him a conveyance in two or three hours for the opposite shore of Albion, which he may also gaze upon daily whilst walking on the pier.'

The 'demon ennui' is surely an oblique reference to the mindset of someone displaced from their homeland by circumstances either of their own making or beyond their control. Pointing out the relative brevity of the crossing and the fact the impecunious Briton could see the shore of his or her homeland from the pier at Calais might also betray a little of Albany's own feelings towards people escaping their debts. His snark was the least of their problems, though: not only was the Channel a barrier, most of them couldn't venture any further than Calais even if they wanted to.

James Albany may have found it a cinch obtaining a passport at London's French embassy, but that granted the holder admission only to their port of entry – to travel further one needed to apply for and obtain more paperwork once one had reached the other side of the Channel. The contrast between Britain and France in this respect couldn't have been more pronounced: the traveller to England needed no passport or indeed any papers whatsoever – he or she just disembarked and that was that, they could stay as long as they wanted. France was considerably tighter when it came to controlling who came into the country, something at which the British rolled their eyes, seeing it as a classic example of the French love for petty bureaucracy: separate papers and permissions were needed first to enter the country and then to move around inside it. Hence, if you arrived at Calais even with a passport you required the permission and stamp of the local *directeur* – something you needed evidence of means to obtain – before you could leave the walled town and continue your journey. Thus the walled city of Calais – not to mention other port towns along the Channel coast of France – through the eighteenth and most of the nineteenth centuries was effectively a laid back, civilised debtors' prison for Brits, a holding camp for the financially embarrassed.

Add to the debtors those couples running away together to escape the inconvenience of aggrieved spouses (Percy Bysshe Shelley and Mary Wollestonecraft Godwin showed up at Dessin's in 1814, requesting Sterne's room) and the unfortunate Brits caught in France when the Peace of Amiens came to an abrupt end in 1803 – some of whom ended up stranded for more than a decade until the French border opened again – and Calais must have had quite the atmosphere during the first half of the nineteenth century, not least given that some of those refugees were among the most famous names of the age, two of which in particular still ring familiar today.

When Horatio Nelson died at Trafalgar in 1805 he had in his last words begged that the state 'take care of Lady Hamilton', and in the codicil to his will that he wrote on the morning of the battle he requested the nation award Emma, the mother of their illegitimate daughter Horatia, 'ample provision to maintain her rank in life'. These instructions were ignored and Emma received a much lower level of financial support than might have been expected for the partner of a national hero. With Nelson and Emma not married, the admiral having left his wife for her and set up home with Emma and her husband, moral mores trumped propriety, the woman was considered the villain of the piece and Emma was quietly written out of the post-Trafalgar narrative, not even invited to Nelson's enormous state occasion of a funeral. By the end of the decade she found herself in debt to the tune of £15,000, and in 1812 was required under the laws of the time regarding debtors to move to an area within the boundaries of the King's Bench in Southwark. Officially she was in prison, but well-to-do debtors were spared the cells and permitted to live effectively under house arrest at an appointed place within a three-mile radius of the debtors' court.

The publication of Nelson's love letters in 1814 erased what public support remained for the blacksmith's daughter who was Nelson's widow in all but name, most assuming she had sold the letters whereas it was a servant who had illicitly transcribed them and passed them to a publisher. Emma realised the only chance of retrieving any kind of life for herself was to leave the country altogether, practically out of the question as long as the French border remained closed. In July 1814, however, with cross-Channel traffic resuming after an eleven-year hiatus, she managed to escape the harassment of her creditors and embark on a flit to Calais with Horatia on a private ship that left in the dead of night from the quay by the Tower of London.

After a traumatically choppy three-day crossing Emma took rooms at Dessin's for a few weeks before the expense dictated she take a less sumptuous abode, a dingy apartment on a dingy street rented from a Monsieur Damas in the Rue Française. Emma was possibly already an alcoholic by the time she reached Calais, and the grimness of her situation led her to spend most of her final days in bed drinking cheap wine and taking laudanum in an effort to obliterate the misery of her existence. She died on 15 January 1815 at the age of forty-nine, and according to some accounts the captains of the British ships in Calais harbour all attended her funeral out of respect for Nelson. The funeral was quite a contrast to her erstwhile partner's: a black petticoat stitched to a white curtain was hurriedly run up to serve as a pall over her coffin.

A year after Emma Hamilton's death, the man George IV had been expecting to see in the crowd at his wet-trousered procession to Dessin's arrived in the form of Beau Brummell. Once the pinnacle of English dandyism, by 1816 the former officer of the Light Dragoons had both alienated the then Prince of Wales (the future George IV) and run up gambling debts so staggering that skipping the country seemed his only pertinent course of action. On 16 May his carriage made it to Dover without being spotted and was loaded onto a hired boat destined for Calais where, with no passport to permit onward travel, he would remain for the next fourteen years. He relied on the support of friends back home to keep him afloat even in a city as cheap to live in as Calais, and like Emma Hamilton first took lodgings at Dessin's, a stay funded by handing over his swanky carriage to the hotel before moving to lodgings above a bookshop on Rue Royale almost opposite, maintaining the affectation of having his meals sent over from the hotel kitchens.

Brummell became a regular sight on the streets of Calais, walking his poodles and inviting notable English travellers to cross the road from Dessin's and join him for afternoon tea.

By 1820 the cash supply from even his most loyal benefactors had all but dried up, meaning Brummell had to find other ways of supporting himself. He made sure he was seen in the summerhouse at Dessin's writing furiously, observers guessing, as he hoped they would, that he was writing his memoirs. The inevitably scandalous contents of such a book would have made it a nailed-on bestseller that would change its author's fortunes as soon as it hit the shelves. Another way of changing those fortunes, he surmised, without having to go to the trouble of actually writing the book, was to send letters to certain establishment figures back home who might have been willing to pay considerable hush money to keep certain anecdotes out of any mooted memoir from across the Channel, not least his old friend and partner at the gambling tables the Prince Regent, who later that year became George IV.

It's unlikely that Brummell would have sent any threats of blackmail, however veiled, however couched in a jocular we-had-some-laughs-didn't-we bonhomie, to his old friend, but the news that George would be passing through Calais must have had him all of a tizz. A reunion might have been propitious to his future, maybe even resulting in some kind of royal appointment, a diplomatic post perhaps, but equally such a meeting would only throw into sharp focus Brummell's much reduced circumstances, hard on both bulwarks of his character: pride and ego.

Whereas he was usually a noted sight perambulating the city walls as passengers arrived on the packets from England – enjoying at least a hint of the fame he'd taken as his right on the streets of London – Brummell stayed away from George's soggy arrival. Nor was he among the crowds lining the route of the king's procession to Dessin's. The closest the pair came to meeting was when Brummell popped into Dessin's and wrote his name in the hotel guestbook, the equivalent of leaving a calling card. If George even heard about it he didn't act upon it, and the pair would never meet again. When the king, climbing into his carriage, left the hotel for the next leg of his journey, and as the horses strained the royal convoy into juddering movement, he reportedly observed, 'I leave Calais and I have not seen Brummell.'

It was a decade before Brummell's fortunes looked up, and by the time they did George was dead. In 1830 William IV appointed Brummell to the British consulship at Caen and he couldn't leave Calais quick enough, departing in such a rush he left behind debts amounting to nearly 12,000 francs, much of which was for meals he'd taken on account at Dessin's. Two years later the post was abolished – at Brummell's suggestion, in a catastrophic attempt to secure a higher position – and he was destitute again. His last years were spent in greater debt and suffering the creeping effects of syphilis, until his death in 1840 in an asylum outside Caen.

The debtors' loophole of Calais was finally closed by an extradition Act passed in 1870 and a treaty signed by Britain and France in 1876. By then Dessin's was in terminal decline, not least because improvements in rail services had meant travellers were no longer detained at Calais waiting for coaches. The hotel was sold to the town in 1860 – the name Dessin's was attached to another establishment nearby – and turned into a museum, displaying, according to one visitor, 'Indian boats, skeletons of birds and fishes, arrows, pictures etc.'. The building is long gone now, demolished in 1880, and an austere, rectangular, flat-roofed school built in its place, a construction that horrified travellers who remembered the old hotel.

The modern building that replaced the school contains no

hint of the brimming chapters of Anglo-French history that were written here. It's almost impossible to imagine now the rambling pile with fountains playing in its twin courtyards, the impressive gardens, the theatre regarded as one of the finest outside Paris, the dining room with Sterne's portrait over the fireplace – not to mention the curious travellers creaking along the corridors counting the door numbers until they reached 31 and standing agape at the sign proclaiming it as Sterne's room. And it's almost impossible to picture the weary seasick travellers falling through doors into the dark wood-panelled rooms and crawling into the clean bedlinen before even drawing a fire in the grate, or the eloping or adulterous couples, no need at last for subterfuge and secrecy, bundling each other onto the bed, giddy with excitement and lightheaded at their new freedom.

There are worse buildings to look at in Calais than the smart, modern red-brick construction that stands where kings and literary giants once passed, and there's probably something appropriate in the fact that along the extent of Dessin's former frontage are a pharmacy, a bank and a restaurant. But to me, one small plaque commemorating a brief stay by a Belgian monarch seems a woefully understated memorial. Dessin's was one of the most famous establishments in Europe for more than a century. If it weren't for the apparent determination of Belgians to commemorate their king, there would be no marker at all.

Wandering further along the Rue Royale I happened upon a restaurant with the appropriate name Ancienne Histoire. As I read the menu on the board outside, a man walked up, pushed the door open, paused, beamed and said to me, '*C'est très bon!*' Taking as I do everything anyone tells me at face value, this complete stranger's recommendation was enough for me to follow him through the door, where he was proved entirely correct. The dessert alone, Minestrone de Fraises, was one of which Dessin himself would have been proud. If the old place can't be commemorated in stone, well, a spectacular dessert nearby will just have to do instead.

I was up early the next morning, a hazy, warm autumnal day, to undertake a now familiar ritual in an unfamiliar place. The beach at Calais is one of fine sand that shelves so gently that at low tide the sea seems to go out for miles. When I'd walked from the hotel round the marina and past the old Second World War fortifications I'd found myself emerging onto a beach bathed in a strong bleaching light, where an expanse of pale-grey sand swept elegantly away to the west towards the cliffs. There was a clutch of small, angular, odd-shaped buildings where road became beach, most of which were empty, but one was a small café, outside which was a four-foot smiling plastic dolphin standing upright on its tail with rubbish sticking out of its mouth, plastic bags full of food containers and the odd plastic bottle. On its stomach someone had written '*poubelle*' in marker pen, apparently in order to confirm that the dolphin had a practical function as a dustbin and wasn't an artist's comment on what it is we're doing to our seas.

The sand was so smooth it could have been rolled. A decent breeze from the west made loose grains shimmer across the surface in constantly shifting ethereal shapes like the ghosts of past beaches. The sea was a luminous green with hints of white caps beyond the protection of the pier, and it looked very inviting.

The tide was halfway in when I got to the shore and stripped down to my red swimming shorts. I made my way to where tiny waves were smearing over the sand and each other, the water beautifully clear, and began to walk in. There were a couple of distant figures on the sand away to the west but mine were the only footprints I could see, stretching back to where I'd left my stuff in a little pile. I walked. And walked. And walked. Where at home I'd have been up to my neck within about five yards, so gently did the sand slope here that it was barely halfway up my shins.

I kept going, aware of the pier to my right and aware there were people on it, walking or fishing, or waiting to watch the ferry on the far side that was churning black smoke into the sky in preparation for its departure, and still I kept walking. The water reached over my knees now but showed no inclination to go much higher. By the time the hem of my shorts got wet I felt like I was halfway to Dover. I was now so far out, heads were turning on the pier, yet not even half of me was wet. I sloshed on, half expecting to reach a sudden lip and disappear like Oliver Hardy crossing the stream in *Way Out West*, but it didn't happen. The water reached the waistband of my shorts and people on the pier were nudging each other and pointing. Judging when to transform walking into swimming wasn't usually an issue for me but I knew I had to do it soon, or someone might presume I was staging some kind of low-wattage suicide attempt and hurl themselves from the jetty to save me. To give the impression I was in deeper than I was, I adopted a kind of knees-bent strut like Groucho Marx prowling a hotel corridor in an effort to kid myself and everyone else that I knew what I was at.

Eventually I arrived at a point deep enough that I could sink to my shoulders. From this position I leaned forward and launched into an attempted swim but my feet struck sand with the first kick and I succeeded only in planting my stupid face into the sea as my knees hit the bottom. Some undignified huffing and puffing and a couple of mouthfuls of sandy sea water later I was sloshing shamefacedly back to the beach.

It's probably no surprise that 'Name Your Favourite Statue Outside a Town Hall' isn't a smash-hit, meme-heavy game sweeping the nation's social media. Statues outside town halls are rarely memorable, generally portraying worthy bureaucrats who even when you read the name on the plaque don't leave you much the wiser. There's probably a foundry somewhere with a corner of its warehouse given over to identical statues of a man in Victorian clothing, one hand inside his frock coat, the other holding some papers as he gazes sightlessly into the distance through soulless eyes. Need a statue outside your town hall? Here's your guy – all you need to add is the plaque because, let's face it, nobody knows what he looked like.

In the unlikely event I do end up participating in a game of 'Name Your Favourite Statue Outside a Town Hall ', however, I have both an immediate answer and the winning hand. Calais's town hall is eye-catching enough thanks mainly to its blingy tower, which looks like Big Ben's in Westminster would if it was going to a Dame Edna Everage-themed party. In the middle of the cobbled space in front of the building is a patch of immaculately manicured grass and flowers, the centrepiece of which is a sculpture in aged-green bronze by Auguste Rodin called *Les Bourgeois de Calais*, The Burghers of Calais.

Edward III had been feeling pretty chipper in the summer of 1346. He'd landed his army in France on the Cotentin Peninsula in July and proceeded to clatter his way towards Paris, burning this town and sacking that in the customary manner of the Hundred Years War. This was a tactic called *chevauchée*, which despite having a name sounding agreeably like a sneeze was a ruthless campaign of destruction designed to weaken an enemy country by battering, burning and bludgeoning it into submission. So it wasn't the most subtle of military manoeuvres, but it was one at which Edward excelled, smashing his way south until, within sight of Paris, he turned northward hoping to combine forces with a Flemish army making its way into France from Flanders for some mutual *chevauchée* action. Unfortunately for Edward the Flemish turned round and went home before they could meet, leaving his forces more exposed than he'd have liked on a hill outside Crécy and facing a French force comprising far greater numbers, who attacked the English.

Thanks in large part to the skill of their longbowmen Edward's army pulled off an unlikely and famous victory after which, giddy with triumphant delight, he gathered his men and set off to lay siege to Calais to round off a successful sortie across the Channel.

Edward arrived outside the walls of Calais on 4 September and made it plain they were in for the long haul if necessary, their camp to the west of the town even boasting its own weekly market. Calais was well fortified and sat at the heart of boggy terrain that provided an extra layer of defence. Furthermore, the town could be resupplied relatively easily from the sea. Edward could have picked simpler places to besiege, but Calais was not only strategically important by its very location, it also represented easy access to his Flemish allies, which had beneficial implications for trade.

The Siege of Calais would prove hugely expensive to English coffers – the army's provisions alone occupied more than 800 ships and drew in victuals from across the south of England and Wales. It was also costly to the French king Philip VI, who in an effort to save money had disbanded his army after Crécy, assuming Edward was on his way home. As winter came on, Edward too was forced to reduce the size of his army, and numbers weren't helped by an outbreak of dysentery in the camp. Despite this he tightened his grip around the town, establishing near-total control of the harbour entrance, while Philip's desperate attempts to reconvene his forces left the French in disarray and the citizens of Calais in dire straits. In June 1347 the town's governor Jean de Vienne wrote to Philip informing him that conditions were so bad the besieged people were on the point of cannibalism. On 3 August Calais surrendered, and Edward prepared to expel the French population and replace it with an English one.

According to the medieval historian Jean Froissart, who was contemporary with the events even if he didn't witness them directly, Edward, having endured such a long stand-off, was furious with the Calesians for offering such a determined defence. His first instinct was to massacre the lot of them – part punishment, part freeing up the town so as to populate it with English people. Instead he offered to spare the people if six of their leaders would approach him in penitent dress, bare-headed and barefoot, bearing the keys to both the city and its castle, and surrender themselves to him wearing nooses around their necks.

'With these six,' said Edward, 'I will do as I please.'

First to volunteer was Eustache de Saint Pierre, the town's richest man (at least, before the siege). Five other town notables stepped forward and the six made their way to the gates of the city in the manner ordered by Edward. Once outside the walls they walked into the heart of the English camp where Edward waited to receive them flanked by his lords and knights.

The men knelt before him and asked for mercy, but Edward, still puce with anger at the town the men represented, ordered their immediate beheading. His entourage pleaded with him to show clemency but he was adamant: these six men had to die, and the town was lucky it wasn't all of them. It took an appeal from his heavily pregnant wife Philippa of Hainault to soften his stance, until the king lost patience, stood up, passed the burghers of Calais into her charge, and stomped off. Philippa took the six to her tent, fed and clothed them, and sent them back to the town.

By the 1880s Calais's success as a centre of the lace industry saw the town expand until it merged officially with nearby St Pierre. To mark the occasion the authorities announced both a new town hall and a call for proposals for an appropriate memorial to the sacrifice of the burghers of Calais. One proposal came from Auguste Rodin, arguably the father of modern sculpture. His first submission, in the form of a clay model, was in the traditional style, the six figures striding confidently to their destiny with their heads held high and with Eustache de St Pierre leading, and was enough to win him the commission. The second phase of the process was to present a model one-third the size of the final version, which when it arrived was substantially different from the outline the council had seen. Far from the heroic group of lads marching nobly to their doom, this was instead a gathering that displayed the combination of fear, dignity, heroism, doubt, stoicism, disbelief and anguish that six men in their situation might have displayed. He had also removed the pedestal (on which most public statuary appeared), in order that the citizens of Calais could stand face to face with the burghers. Calais's bureaucrats weren't prepared for this kind of realism, declaring that 'their defeated postures offend our religion'.

Rodin, of course, despite having drastically reworked his original proposal, was having none of this philistine criticism, complaining that the suggestion to revert to the original more conventional design that he had submitted would 'emasculate my work'. The city gave in, and Rodin completed the work in the heart-rending form that stands today. Initially placed in the Parc Richelieu close to the town hall – and, against the artist's wishes, on a pedestal enclosed by railings – it was moved in 1924 to its present position in front of the town hall, where it can be seen and experienced at the closest of quarters.

It really is an extraordinary piece of work, one that would

have lost much of its emotional impact fenced off and raised on a pedestal. Each burgher faces in a different direction and there is no clear leader, emphasising how each man was prepared to make exactly the same sacrifice as the rest, no matter what their relative status might be: this is a distinctly human situation, one in which the subjects are riddled with the most intense emotions imaginable. The six are destined to remain forever grouped together, but Rodin succeeds in emphasising their plight as individuals. There's no camaraderie here, no sense of we're all in this together – not at the moment the sculpture captures. That moment is the one at which each figure realises the full extent of their decision to save the town at the expense of their own life.

The figures are given outsized feet and hands that somehow serve only to emphasise those decisions, especially the distraught figure of Andrieu d'Andres, whose hands are clamped over his head in a posture so overwhelmed it's as if he's about to topple forward onto the ground. Eustache de St Pierre, the man the town council wanted to be displayed heroically as the focus of the memorial, stares at the ground, his facial features sunken after almost a year of starvation and hardship that not even the richest man in Calais could avoid.

It's an astonishingly moving piece of statuary, all the more so for the proximity Rodin allows the viewer. From every angle as you circle the work you see a different face: there is no 'right' angle from which to view it. Among the carefully tended flower beds in the sunshine it's almost like being transported back to that moment when, after enduring almost a year of such hardship that cannibalism was being considered in official circles, these gaunt, tormented, half-starved wretches were given an agonising decision to make, to save the town by sacrificing themselves. Every ounce of that anguish is captured by Rodin.

6

Brighton

I'm guessing that not many people have ever travelled to Brighton expressly to visit the branch of Patisserie Valerie tucked away off a little square on East Street in the heart of the old town. I may well be the only person in history to have walked through the little conservatory area at the front of the building, then up the steps and through the door, and beamed at the sheer delight of being in that particular room in that particular branch of that particular chain of coffee shops.

I am not some kind of travelling Patisserie Valerie enthusiast. I'm sure Valerie is lovely and an absolute dead shot at patisse-ring, but I had a different reason for showing up and sitting down with a coffee and an almond croissant than just sitting down with a coffee and an almond croissant. This Grade II listed building dating back to the eighteenth century had been home to one of the great Channel characters. Not only that, I'd boldly suggest that this particular resident spent more time in the Channel itself than anyone ever before or since. Her name was Martha Gunn, she was the famous Brighton dipper and I was sitting in her living room.

During the second half of the century Martha was one of the most famous women in the country. She was no society beauty, she wrote no novels or poems, she couldn't act or sing, wasn't boffing anyone she shouldn't have been, had no say in the affairs of state, yet Martha had the ear of princes, and her company and assistance were sought by some of the grandest women in the land. She appeared in popular prints like the one the British Museum holds from 1794, a time when the perceived threat of invasion was about as high as it's ever been. Called *French Invasion, or, Brighton in a Bustle*, it depicts French soldiers trying to land on Brighton beach and being set about by locals, the furthest forward of which, standing on a couple of French soldiers and setting about a third with a stick, is Martha Gunn.

Martha was one of the most popular figures ever to be immortalised on a toby jug and today originals featuring her likeness change hands for hundreds of pounds. The Prince of Wales, the future George IV, adored her and granted her free access to his kitchens at the Royal Pavilion whenever she pleased. When Martha Gunn died in May 1815 at the age of eighty-eight, Brighton came to a virtual standstill. Impressive stuff, considering what she did to earn that level of fame: stand in the sea, day in, day out, shoving well-to-do women's heads under the water. For around seventy years Martha spent her days up to her chest in the Channel off Brighton beach, as bathing carriage after bathing carriage trundled and sloshed towards her. When it was in place she would pull open the door, take hold of the woman inside, lift her down, immerse her beneath the waves, swing her around in the water a bit then assist her back into the carriage, where maids awaited her with dry clothes. The door would then be closed and the carriage would trundle and slosh its way back up the shingle and Martha would wait for her next client.

Brighton made Martha Gunn, and in a sense Martha Gunn helped to make Brighton, as what had been a ramshackle fishing village called Brighthelmstone was transformed into the most fashionable town in England. Brighton retains an agreeable mixture of the rough and the refined, its grandeur faded yet somehow simultaneously spruce. It's a Mecca for hipsters yet manages to avoid being insufferably pleased with itself in the manner of certain parts of, say, east London. Brighton's cool is raffishly frayed and retains a commendable sense of self-awareness. The immaculate beards and carefully distressed plaid shirts mix easily with the old lady in a deckchair on the pier, wrapped up in her overcoat and munching through a bag of toffees. Waitrose and Poundland are comfortable neighbours in Brighton, a place where they still eat chips out of a newspaper and it's the *Guardian*. It's a town not entirely free of affectations, but it knows it. It also knows that it's seen just about everything and doesn't really have to try to impress any more.

In the very early eighteenth century, however, Brighthelmstone wasn't really up to much. Little more than a bundle of cottages, a church and an inn, it was a prosperous fishing village until at the turn of that century it began to be flattened regularly by storms, notably the great storm that destroyed most of the southern half of the country in 1703. In his account of that particular tempest Daniel Defoe noted how the village suffered such extensive damage it looked as if it had been bombarded by artillery. Two years later another storm destroyed a number of buildings and buried the wreckage beneath shingle. The population dipped so dramatically, a couple more storms and there was a fair-to-middling chance the place would have been abandoned altogether.

Then a physician named Richard Russell began the process that turned Brighthelmstone into Brighton and transformed the town's fortunes.

The son of a London bookseller, Russell was born in 1687 and educated at Westminster before taking a medical degree in Reims. By 1725 he had his own practice at Lewes in Sussex, where he began to explore theories he was developing regarding the health benefits of the sea. Not just bathing but, well, dunking a tankard in it and necking the contents. From the 1730s he began sending patients to his nearest coastal settlement, Brighthelmstone, with instructions to bathe in the sea every morning and drink a pint of sea water at least once a day. That's drink. A pint. Of sea water. A day.

For all the vomiting and other bodily expulsions the treatment triggered (not to mention the raging thirst that he insisted would diminish the more sea water you drank), Russell's theories took off to such an extent that he was able to build himself a grand house with attached consulting rooms right on the front, almost opposite where the Palace Pier stands today, and published a book advocating sea water treatment for all sorts of agues and conditions. *De Tabe Glandulari, sive de Usu Aquae Marinae in Morbis Glandularum Dissertatio* was its snappy title, rendered into English as *Glandular Diseases, or a Dissertation on the Use of Sea-Water in the Affections of the Glands*, a publication successful enough to land him election to the Royal Society. So popular was his hypothesis that in the 1750s bottles of Brighthelmstone sea water were marketed and sold in London and beyond for those in need of a good purge and thirst but too far from Brighthelmstone to indulge.

'A pint will be found commonly sufficient in grown persons to give three or four smart stools,' he wrote, not referring to furniture, and described how one of his patients drank twenty-five *gallons* of Channel water in one intensive period of treatment that cured her skin complaints and internal disorders in one salty hit. Russell was delighted; it was a triumphant one in the eye for his many doubters.

'But if the drinking of this immense quantity of sea water and its consequences, the restoring and amending of the patient's health, do not sufficiently declare and evince how powerful and how harmless at the same time this medicine is,' he yah-boo-sucked, 'I doubt whether any argument can be found sufficient to overcome the prejudices of mankind, however inconsiderately they may have been taken up.'

As well as launching spumes of briny vomit Russell's treatments effectively launched the Brighton we know today, as the well-to-do began arriving in droves for the benefit of their health. On the site of his house, now the Royal Albion Hotel, there's a plaque on the wall marking the spot that bears the frankly lovely inscription, 'If you seek his monument, look around.' It's lifted verbatim from Sir Christopher Wren's tomb in St Paul's Cathedral, but is no less appropriate for it.

For all the town's increasing popularity its facilities were slow to catch up. The only places to stay were cottages where the fishing families were prepared to budge up for a few coins; there was no paving in the streets, which weren't lit at night, and there wasn't even a local magistrate until the 1770s. There wasn't much going on, either. In 1736 a clergyman named William Clarke visited the town to bathe and wrote to his wife, 'My morning business is bathing in the sea and then buying fish.' His evenings, he gushed, were spent 'counting the ships out in the road and the boats that are trawling' – the hedonistic old maniac. There was little increase in the propensity for fun as the century progressed, but with most visitors knocking back bottles of sea water like there was no tomorrow they'd have been too busy mopping up their purges to notice the place wasn't exactly a cathedral to pleasure. Indeed, not only was Brighthelmstone boring, it was also a dump.

'As to the lodgings in this place the best are most execrable,' wrote one visitor in 1763. An elderly Dr Samuel Johnson turned up thirteen years later, writing to Boswell that 'the place was very dull and I was not well', later telling Hester Thrale, 'It is a country so truly desolate that if one had a mind to hang oneself in desperation at being obliged to live there it would be difficult to find a tree on which to fasten a rope.'

The crowds increased, nonetheless, thanks to the most preposterous medical claims. In 1768 a Dr John Awsiter, an admirer of Russell, wrote *Thoughts on Brighthelmstone* in which he evangelised the health benefits of sea bathing as far as claiming the Channel waters could cure cancer. 'It may not be amiss to observe that by keeping [the tumours] wet with cloths dipped in our English mineral waters the malignity of them has been destroyed and cures performed,' he wrote, 'but I am inclined to think sea water more effectual.' He blurted out similarly quackish nonsense regarding infertility, which in those days was of course blamed entirely on women. 'In cases of barrenness I look upon sea water to stand before all other remedies,' he claimed, advising childless women to head for the south coast and fling themselves into the sea while gulping down as much of it as they could. Sensationalist cobblers it may have been – Russell had even claimed deafness could be cured by sticking one's head in the Channel – but it brought people to Brighton in droves.

By this time Martha Gunn was already a dipper. From an old fishing family, she'd begun assisting people in the water in her teens during the 1740s. Luckily for the fortunes of the local folk, it was not done to be seen frolicking in the sea in one's bloomers: bathing carriages were required to trundle the bather a sufficient distance from the shore for him or her to drop into the water away from prying eyes. Another advantage for the locals was that hardly anyone in the eighteenth century could swim, opening up whole new careers for sea-savvy Brighthelmstonians.

For some reason women were 'dipped' while men were 'bathed' (the word 'bathing' applied to women was presumably loaded with too many erotic overtones for men to cope with without going all unnecessary, even while purging), and Martha Gunn was the queen of the dippers from an early age.

Born Martha Killick, she married Stephen Gunn in 1758 and somehow, given she was permanently clothed in enough layers of insulation against the Channel cold to make her as wide as she was tall, had eight children. Mr Gunn's ardour must have been quite something, as it would have taken Martha a good hour to remove enough clothing to facilitate even the faintest possibility of conception.

Martha wasn't the only dipper: while she was easily the most popular and most famous, there was a group of them who protected their work fiercely, as did the bathers who assisted the men. Martha's male equivalent was a bather called John 'Smoaker' Miles, another to win the affections of the Prince of Wales despite once having to grab the enthusiastic royal by the ear as he thrashed about in the sea, explaining, 'I aren't going to let the king hang me for lettin' the Prince of Wales drown hisself, not I, to please nobody, I can tell 'e.' When Miles died in 1784 the prince named the Smoaker Stakes at Brighton races in his honour.

The dippers and bathers limited the number of carriages in use on the beach to ensure the demand for their services was maintained, and after their day's work was done they'd go to Castle Square and wait for the coaches from London to pull in, giving out their cards to the new arrivals. So popular was Martha that footmen sometimes even came to blows in order to secure her services for their charge.

Sitting in Martha's house, I tried to picture her coming in from her day in the Channel, her clothes dripping wet, and gradually removing the layers to hang them in front of the fire. A local myth has it that the bulge in East Street forming the little piazza where her house stood was once an inlet of the sea, so it's possible the Channel waters even washed up against her doorstep.

She may have been one of the great characters of the English

Channel, but she left behind no written record; only a few anecdotes have been handed down, to add character to the prints and toby jugs. I did find one record of a conversation that took place a few months before her death in a book called *The Observant Pedestrian Mounted, or, A Donkey Trip to Brighton*, dated 1815. It's billed on its title page as a 'comic sentimental novel', but the exchange is convincing enough to have surely been based on a real encounter.

The author, whose name isn't listed, is out for a stroll one evening and recognises Martha as she hands out cards close to the coaching office. 'I don't bathe cos I a'nt so strong as I used to be so I superintend on the beach, for I'm up before all of them,' she says. 'You may always find me, up with my pitcher, on the same spot every day before six.'

Martha reveals that she'll be eighty-nine 'come next Christmas pudding', and 'though I've lost my teeth I can mumble it with as good a relish and hearty an appetite as anybody'.

When the author says how glad he is to hear she's in good health, because Brighton wouldn't be the same without her, Martha replies: 'Oh I don't know, it's likely to do without me some day. But while I've health and life I must be bustling around my old friends and benefactors. I think I ought to be very proud, for I've as many bows from man, woman or child as the Prince himself. Aye, I do believe the very dogs of this town know me.'

I walked out of the café and headed up the hill. Possibly a matter of weeks after that conversation Martha died and was buried at St Nicolas's Church beside her husband, who had died two years earlier. Her funeral procession would have taken the same route, passing close to the Royal Pavilion and up towards what was probably the highest point in the town back then, and it didn't take me long to locate her final resting place, high above the Channel whose salty water after seventy years of constant immersion must have seeped right into her bones. The headstone is well kept and the grave surrounded by an iron railing of more recent vintage than 1815, suggesting this tremendous old Brightonian has not been forgotten by the town. As well as recording her as 'peculiarly distinguished as a bather in this town' the stone also commemorates four of her children, who predeceased her in the space of four years between 1784 and 1788 at ages ranging from thirteen to thirty.

While foraging for Martha in the archives I found a notice in an old issue of the *Hampshire Chronicle* from 1804 that would have served as appropriate inspiration for her epitaph. 'Martha Gunn has full employment daily,' it read, 'and many a blushing beauty by the Venerable Priestess of the Bath is soused over the head and ears in the pickling-tub of Neptune.'

The following morning I too was soused over the head and ears in the pickling tub of Neptune but not, alas, by Martha Gunn. I was staying in a flat on the tenth floor of Embassy Court, an Art Deco block of seafront apartments built in 1935 on the border between Brighton and Hove. My sister had moved in there as a student during the early 1990s, when the place was falling to pieces after decades of neglect and was home to a motley collection of squatters, artists, junkies and, well, my sister. These days the building is a little more reminiscent of its glory years when the likes of Diana Dors and Laurence Olivier had flats there. The ground floor featured a bank and a restaurant for the residents' exclusive use, and there was a roof garden for cocktail parties. Keith Waterhouse, who described Brighton as looking like it was permanently helping police with their inquiries, lived for a while in the flat directly below the one in which I was staying.

I was up early, looking out across the rusty skeletal birthday cake of what remains of the old West Pier to the twinkly lights of the Palace Pier beyond, pin sharp in the dawn twilight where the sun would be seen rising if it wasn't a blustery, cloudy late-November morning. Down on the beach I noticed a group of people about a dozen strong, all in swimwear, with a fire lit on the shingle, running at full tilt into the breakers that thundered onto the stones. I stepped out onto the balcony, the wind slopping my tea over the sides of the mug, to hear among the roar of the waves below the occasional yelp of a swimmer. This was impressive stuff – the waves looked enormous even from ten floors up, white foam stretching a good thirty yards or more out from the shoreline, making the dark spots of the swimmers' heads easier to pick out as they were lifted and dropped by the power of the Channel. It wasn't long before they were staggering out of the surf, whooping and laughing with the disbelief of exhilaration, and towelling off around the fire.

A few minutes later I was in my shorts, in my fleecy swim hoodie and in the lift. After crossing the road and passing the beachfront café where the light seeping from the edges of its shutters showed it was about to open, I scrunched over the shingle and down to the sea to find conditions very different from what I was used to. There were no Goodwin Sands here to take the sting out of a big sea, and while what I was witnessing wasn't by any means the full wrath of the Channel this was a far stroppier sea than the one in which I swam every morning. The waves crashing on the pebbles roared at me as I unzipped my hoodie and dropped it onto the beach, anchoring it with a couple of handfuls of stones. The wind had the hem of my shorts fluttering like a child's sandcastle flag as I walked towards the sea with as much purpose as I could muster, even though I didn't really fancy this at all. A wave crashed in front of me, the spray flying up into my chest and pebbles strafing my shins.

I had to time this right. When the next wave hit I ran through the spume, hoping to reach the following one before it broke. I wasn't quite quick enough and it almost knocked me onto the pebbles. I pirouetted ungracefully and needed to put a hand down to stop myself going over. Standing up, I took the next wave in the small of the back, then turned and went a few paces further into the sea. I was deep enough now not to catch the waves as they broke, and jumped to go up with the next one, the tip of it white with foam as it prepared to crash down. Off the bottom now, I tried a few strokes but the force of the water rendered my efforts as not so much swimming as flailing.

The sea was dark, almost black, under the threatening November sky, the wind whipped spray into my face and when I tasted salt water I realised I was laughing. I gave up thoughts of swimming and rode the waves for as long as I could take it, deafened by their crashing and by the roaring of the wind. The sea lifted my legs under me: the Channel was going to deposit me back on the shore whether I liked it or not. I thrashed at a couple of strokes but as soon as my feet touched the bottom they were lifted again by the biggest wave yet. I was thrown forward and my feet hit the stones as my top half was swept towards the beach, and before I knew it I was tumbling out of control in the foam. My mouth filled with water, my nostrils stung, my shoulder struck pebbles, my feet came over the top and I was being rolled out of the water and up the beach, still propelled by the sea. The next thing I knew I was on my hands and knees, blinking away the salt water running from my hair into my eyes, watching the foam wash around its parabola and then be sucked back off the beach in a cascade of pebbles that tumbled towards me. I stood up and the next wave came in waist height and I was on all fours again before dragging myself up the beach to the warmth and safety of my fleecy hoodie. My skin sang, my heart pumped, I could feel every nerve in my body. A small cut on my left knee aside, I was unscathed.

The café was open by the time I exited the beach. I'd probably swallowed enough sea water to expect a decent Russellian purge before the day was out, but I chased it down with a polystyrene cup of tea and an egg and bacon roll, the paper plate bending around it in the wind. I had no fire to warm by on the beach; instead I went back across the road, dripped sea water and egg yolk in the lift, found the flat, walked out onto the balcony again and looked down at the breakers, still roaring away, no trace at all of my encounter. The Channel had played with me for a bit and then thrown me out, back to where I belonged.

'Everything here is in motion – coming or going,' wrote William Hazlitt of Brighton before departing for the continent in 1824, and right at that moment, my senses still ringing, my skin still fizzing and rough with the salt, my breath still coming fast, I sort of knew what he meant.

Many writers have composed verse and prose about the Channel, but for me if this stretch of water has an immortal laureate it is Charlotte Smith who, in October 1784, stood outside the Ship Inn on Brighton seafront waiting for the weather to abate so she could cross to Dieppe for the next chapter of a tumultuous life. As she loitered, pregnant, with five of her children and their belongings piled up next to them, a large part of her must have been hoping the weather would never change, the wind would continue to howl from the south-west, and she'd never have to make the crossing.

Charlotte Smith was one of the key figures of early Romanticism but, being a woman, she's never received the credit she was due. Wordsworth was a fan; Coleridge and Southey raved about her work. During the late eighteenth century she helped to relaunch the sonnet as a popular poetic form as well as publishing ten bestselling novels that became important influences on Jane Austen, among others. Despite these impressive literary chops there's a good chance you've never heard of Charlotte Smith. Her popularity was on the wane even before her death in 1806 and by the mid nineteenth century she'd practically dropped out of the national literary narrative altogether. There are editions of her poetry in circulation today if you look hard enough, but her novels are so long out of print that copies available online go for hundreds of pounds.

Smith was a woman out of her time. She made no secret that she wrote not from some pure artistic impulse but to support herself and her children, especially once she'd left a shifty, feckless, abusive husband to become an independent single mother making her own living. 'I loved novels no more than a grocer loves figs,' she wrote towards the end of her life: she possessed the kind of straight-talking independence that in the eighteenth century made influential men nervous and left posterity not quite sure what to do with her. She'd spent the duration of her marriage trying alternately to negotiate with and to fend off her husband's creditors, joining him in a debtors' prison for a while and going along with his ill-fated plan to move the family into a decrepit rural chateau in northern France in an attempt to escape his debts.

Once she'd left him, Smith would spend the rest of her life trying to secure the inheritance promised to her and her children by her well-meaning father-in-law, a man who didn't want to see it frittered and gambled away by his idiot son but whose whimsical will triggered legal wrangling that would outlive Smith herself, and not be resolved for thirty-six years.

Added to having to move her brood – she had twelve children of whom only six would survive her – from house to house in line with her fluctuating finances was the fact that under the prevailing culture of primogeniture her husband was legally entitled to all royalties from her books, even though they were separated and despite the rheumatoid arthritis that made the physical act of writing itself incredibly painful. It's a wonder Charlotte Smith wrote anything at all.

Her novels argued in favour of greater rights for women and shone a light on how women were subjugated by marriage and the patriarchal laws of the day. She became a passionate advocate of the French Revolution; her political opinions fanned out from Europe across the Atlantic – in her American-set novel *The Old Manor House* she dispenses subtle yet hard-hitting criticism of the injustices meted out to Native Americans, as well as knocking lumps off British colonialism and the slave trade. 'Mrs Smith writes like a gentlewoman,' wrote Mary Wollstonecraft. 'If she introduces ladies of quality they are transcribed from life and not the sickly offspring of a distempered imagination that looks up with awe to the sounding distinctions of rank and the gay delights which riches afford.'

For Robert Southey, Smith was 'a woman of genius, good sense, and pleasant manners', a faintly patronising endorsement but a sincere one, while many scholars, particularly the handful of feminist academics who have in recent years striven to achieve greater recognition for Charlotte Smith, regard her as an instigator of Romanticism as instrumental as Wordsworth and Coleridge themselves.

Born into a wealthy family in London in 1749, Charlotte Turner lost her mother at the age of three. The loss of his wife prompted her father to pack his bags and take off on a tour of the world's casinos, spending as if allergic to cash while leaving his three children in the care of an aunt. The enthusiasm with which he frittered away the family's money and assets led to Charlotte's betrothal at the age of fifteen to Benjamin Smith, son of a successful merchant trading with the West Indies and a man nearly a decade her senior; in later life she would write that her father had effectively sold her into 'legal prostitution'.

Benjamin Smith was a dreadful husband and wasn't much

better at being a human. Hopeless with money, openly and serially unfaithful, regularly physically violent towards his wife and always trying to stay one step ahead of his many creditors, he was sent to the King's Bench Prison for debt in 1783 where, as was the custom, his wife joined him. Charlotte spent her days writing letters trying to secure the means to release her husband and, more significantly, composing the verses that would be published to wide acclaim the following year as *Elegiac Sonnets*, running to nine editions in six years – the list of subscribers to the fifth edition alone included the Archbishop of Canterbury, Fanny Burney, the literary bluestocking Elizabeth Carter, William Cowper, William Pitt and Horace Walpole.

After seven months in the debtors' prison Charlotte finally managed to put deals in place to secure her husband's release, and he immediately crossed the Channel, moving into a chateau in the countryside between Dieppe and Rouen, with Charlotte bringing the children to join him after that gale-delayed crossing in 1784. The chateau was large, decrepit, riddled with damp and freezing cold, and that winter the thirty-five year old went into her twelfth labour, giving birth to a boy. She couldn't face any more pregnancies and resolved, in the depths of that French winter as the wind whistled across the fields and through the shutters, that she would earn her own living from writing. When she found a copy of *Manon Lescaut* in the chateau she set about translating it, then sold it to a London publisher, which gave her an income that, combined with her poetry royalties, permitted a return to England.

In 1787 she left Benjamin. 'Tho' infidelity, and with the most despicable objects, had rendered my continuing to live with him wretched long before his debts compelled him to leave England,' she wrote later, 'I could have been contented to reside in the same house with him had his temper not been so capricious and so cruel that my life was not safe.'

In 1789, the same year her novel *Ethelinde* came out, Charlotte moved to Brighton for the benefit of her health, suffering from what was most likely rheumatoid arthritis, and for the radical political discourse she found there in the aftermath of the French Revolution.

Supported by publishers who would try to keep the proceeds of her work away from her husband, Smith became a prolific author in the then-popular Gothic style who published nine novels in as many years to support her family and fund the legal action aimed at securing her father-in-law's promised fortune, a legacy hotly contested by his family. By the time she arrived in Brighton she already had two bestsellers under her belt, not to mention the success of the *Elegiac Sonnets* that she felt lent her a literary respectability her fiction couldn't, even if the latter had brought her widespread fame.

Western Road, running parallel to the shore and linking Brighton and Hove, has always been a busy thoroughfare. A walk along it today produces an aromatic mix of cuisines from all over the world: India, South America, Asia, North Africa and the Middle East. No one knows for sure where exactly Smith lived – her surviving letters specify only a location close to the sea, but it's likely to have been in a lodging house on one of the streets of Georgian houses that lead down to the shore from Western Road.

It was there in November 1791 that she received a visit from an enthusiastic young fan, a meeting that would have far-reaching repercussions for English literature. William Wordsworth was twenty-one years old, unknown and unpublished when he arrived in Brighton that winter, recently graduated from Cambridge and heading for revolutionary France. Wordsworth had long been a fan of Smith's work and his copy of her *Elegiac Sonnets* had been with him since his schooldays at Hawkshead, leaving it well thumbed and extensively annotated. So when he found his crossing to Dieppe delayed by bad weather he checked himself in to Brighton's famous Ship Inn and made inquiries as to the whereabouts of Mrs Smith.

Gales detained him for four days, 'which time must have past [*sic*] in a manner extremely disagreeable if I had not bethought me of introducing myself to Mrs Charlotte Smith,' he wrote. 'She received me in the politest manner and showed me every possible civility.'

Charlotte was recently returned from France herself, having paid a visit to Paris to get a feel for the Revolution – a cross-Channel trip that would inspire her politically radical novel *Desmond*, another bestseller – and Wordsworth would have been rapt to hear her account of the journey. She was also witnessing the steady torrent of refugees arriving on the Channel shore: aristocrats, clergymen and the middle classes, some of them washing up at night in small boats, exhausted, sometimes having rowed the whole way with just the clothes on their backs.

Her meeting with the young poet in the modest drawing room of a Brighton lodging house helped cement Wordsworth's destiny. They would most likely have discussed Smith's *Elegiac Sonnets*, in which not only did she place the writer in nature at a specific time and place, she also developed a more relaxed structure than the traditionally rigid Petrarchian form, an influence that can be easily discerned in Wordsworth's own sonnets. Two years later she would publish her epic poem *The Emigrants*, a beautiful evocation of displacement seen through the eyes of those fleeing the September 1792 violence and murder in Paris, and whose themes and nuances can be identified in Wordsworth's *Lines Composed a Few Miles Above Tintern Abbey*.

The opening passage of *The Emigrants* is one of the most beautiful pieces of writing about the English Channel, evoking its power, its indifference to life, its symbolism as a barrier and its hint that on the other side might lie a haven for the oppressed and marginalised.

> SLOW in the Wintry Morn, the struggling light
> Throws a faint gleam upon the troubled waves;
> Their foaming tops, as they approach the shore
> And the broad surf that never ceasing breaks
> On the innumerous pebbles, catch the beams
> Of the pale Sun, that with reluctance gives
> To this cold northern Isle, its shorten'd day.
> Alas! how few the morning wakes to joy!
> How many murmur at oblivious night
> For leaving them so soon; for bearing thus
> Their fancied bliss (the only bliss they taste!),
> On her black wings away!
> Changing the dreams
> That sooth'd their sorrows, for calamities
> (And every day brings its own sad proportion)
> For doubts, diseases, abject dread of Death,
> And faithless friends, and fame and fortune lost;
> Fancied or real wants; and wounded pride,
> That views the day star, but to curse his beams.

Walking today along Brighton seafront towards what's now called the Old Ship, it's possible to imagine Wordsworth's excitement as he made the same journey back to his rooms, all breathless delight amid the boom of the waves and the crunch of feet on shingle. Inside his coat he carried letters of introduction from Smith to political radicals of her acquaintance in Paris, a generous bestowal on a pushy, idealistic young fan who'd arrived unbidden on her doorstep, but an act that made Smith a key political as well as literary influence on the young poet.

Yet Charlotte Smith's writing is good enough to stand alone without having to point at her influence on more famous names. Her independence, openness and political radicalism, toned down though it was in her novels so as not to alienate readers, worked against her, and mainly because of her gender. She revolutionised the sonnet, helped establish the form of the Gothic novel, planted the seeds of the country house as allegory for society in the imagination of Jane Austen, and despite the hardships and obstacles she faced forged a life and career on her own terms.

It's unfair that she's not widely read today. As Wordsworth himself put it, she was 'a lady to whom English verse is under greater obligations than are likely to be either acknowledged or remembered'.

The Brighton from which Charlotte Smith departed for France was about to change dramatically, and the catalyst was Martha Gunn's royal pal George, Prince of Wales. Richard Russell died in 1759, after which his house was a popular holiday let for the well-to-do, including Henry, Duke of Cumberland, a brother of King George III. The Prince of Wales first visited Brighton in September 1783 to stay with his uncle, remaining for the best part of two weeks which he spent bathing, sailing, hunting, going to the theatre and attending a ball at the Castle Tavern. Away from the hothouse atmosphere of George III's court he could immerse himself in the more relaxed household of his uncle, a renowned shagger who'd infuriated his brother the king by marrying a commoner after a string of dalliances with other men's wives.

The prince had a whale of a time and came back the following year hoping to ease the swollen glands in his neck with some good old-fashioned Russellian bathing. This time he rented a farmhouse looking out onto the Steine, a green area at the heart of the town that the original fishing community had used to dry their nets on. So well did he take to the town that he began a relationship with it that was to last forty years, leaving a mark and an influence on the place that can still be discerned today. His most obvious influence can be seen in the extraordinary Royal Pavilion, which began as a series of extensions to the farmhouse whose lease he'd taken over permanently and developed into what Hazlitt described as the 'collection of stone pumpkins and pepper-boxes' we know today.

As soon as he'd arrived in Brighton George set about acquainting himself with the local ladies. In the eleven days of his first visit, between the hunts, bathes, theatre visits and balls he managed to squire a woman named Perdita Robinson, then discarded her for Mrs Grace Dalrymple Elliott, known as 'Dolly the Tall', with whom he possibly fathered a daughter (three other noblemen came forward to present cases for possible paternity). His conduct in Brighton was, in the words of one sniffy antiquarian history of the town, 'restrained by no considerations of decency'.

As well as the ladies, George surrounded himself with lads, a collection of boisterous lah-di-dahs whose wealth was in direct inversion to their intelligence. One night two of George's Brighton friends frightened the living daylights out of him by sending into his room, while he slept off a night on the claret, a donkey with a set of antlers strapped to its head. Top bants, lads, really. Top, top bants. George himself took to wandering around the town with a gun, shooting chimneys off houses just for the devilment.

At around the time he became infatuated with Brighton he'd also become infatuated with the twice-widowed Maria Fitzherbert, a woman six years his senior and a Roman Catholic to boot. George was determined to marry her despite the Royal Marriages Act saying that any heir to the throne who married a Catholic renounced their claim, as well as another Act passed by his father in response to his uncle's nuptials that barred royals from marrying commoners.

The couple dodged these legislative humbugs with frequent escapes to Brighton, where Martha Gunn could only have endeared herself to them by greeting Fitzherbert as 'Mrs Prince' whenever she loomed above her in the doorway of a bathing carriage.

George's mix of hedonism and nobility, not to mention the delightfully exuberant triumph of fat wallet over taste that is the Royal Pavilion, was largely responsible for the Brighton we know today, a town created in something approaching his own image. It wasn't a deliberate moulding – he clearly loved the place and felt more at home there than he ever did in London, and prince, town and timing were a perfect combination. The people, despite their chimneys being at risk from the royal grapeshot and their women from the royal glad eye, loved him back.

There can be no town or city in the country whose character has been forged to such an extent by one person. Pavilion aside, the architecture still betrays the mark of the Regency, not least in the squares that open out onto the seafront, but it's in the town's personality that his influence is strongest. Brighton remains a haven for the eccentric, the marginalised and the creatively blessed. It's also Britain's most gay-friendly city and has been since the nineteenth century, introducing women-only tea dances in the 1920s and a men-only beach after the Second World War. Brighton has also retained a rough edge, from its gangsters of the early twentieth century as portrayed in Graham Greene's *Brighton Rock* to mods and rockers battering the living daylights out of each other on bank holiday weekends, and there's still a faintly ragged edge to the place today. The architecture critic Ian Nairn said Brighton is what London would be if everyone was removed from the city bar the duchess, the spiv and the cockney, while a guidebook from the late eighteenth century summed it up thus: 'The coast there is like the greater part of its visitors: bold, saucy, intrusive and dangerous.'

There's a permanent sense of performance in Brighton, an air of anticipation, as if you're waiting for the curtain to go up and the show to start, and the show is the town itself. No wonder the outsider has always felt at home here in a way he or she might not elsewhere: Brighton rejoices in the eccentric and the out-of-the-ordinary in a way few places in the world can match.

Close to Martha Gunn's grave, for example, is that of Phoebe Hessel, who died in 1821. Stepney-born as Phoebe Smith, at the age of fifteen she fell in love with a soldier named Sam Golding, and when his regiment was sent to the West Indies, rather than lose him to the Army on learning the Fifth Regiment of Foot was to be posted to the Caribbean, she disguised herself as a man and enlisted. She and Golding became career soldiers, Hessel successfully concealing her gender even after being wounded at the 1745 Battle of Fontenoy during the War of the Austrian Succession. Eventually, when Golding was badly injured and sent to Plymouth to convalesce, Phoebe confided the truth to a colonel's wife who arranged for her discharge and return to England to nurse Golding back to health. He was subsequently invalided out of the Army, freeing the couple to marry at last.

Despite being in their thirties they produced nine children, eight of whom died in infancy (the ninth drowned in Plymouth harbour). When Golding died Phoebe moved to Brighton and married Thomas Hessel, becoming widowed again at the age of eighty and making her living selling fish and vegetables on the streets. At one point, times became particularly hard and she was forced to enter the workhouse, but when George, who like most Brighton residents had heard her story, learned of her plight he ordered her discharge and awarded her an annuity. She spent the rest of her days on the corner of the Steine and Marine Parade selling sweets, pincushions and small toys, and at the town's 1815 celebrations marking the Battle of Waterloo she was seated at the vicar's right hand in honour of being the town's oldest inhabitant at 102. According to her headstone she was 108 years old when she died.

Probably my favourite Brighton eccentric dates from the glory days of the Regency, a man who was clearly vulnerable yet could be at home in Brighton more than anywhere else. Henry Cope was well known around the town at the turn of the nineteenth century for his remarkable appearance. He was known to everyone as the Green Man because he dressed from head to foot in green clothes, powdered his face with green powder and dyed his hair green. All the rooms in his home were painted green and were furnished only with items and objects of the colour green. He ate only green food – vegetables and fruit – in an era when vegetarianism was all but unheard of. He was clearly a man of means, and spent most of his days strolling up and down the front and around town visiting tea shops and the library as if there was nothing remotely unusual about his appearance. When he engaged with others he introduced himself as 'Earl Vernor', a title that didn't exist. There were the usual kind of rumours, that he had suffered a broken heart, even that the green theme had been prompted by a jilting at the anvil altar of Gretna Green. But Henry Cope, the Green Man, was just another Brighton character.

From 1806, however, Cope showed increasing signs that he was probably a little more than purely eccentric. That year he was reported to have sent his manservant to London to source a poulterer who could provide him with a regular supply of green geese. In October there was sniggering from the salacious *Morning Post* at his increasingly odd behaviour, as it suggested that the cut of his green suit was under threat from 'certain gentlemen who would prescribe a *straight waistcoat* to add to his outfit'. Clearly, Cope was suffering some kind of breakdown because at the end of that month it was reported that he had tumbled out of the window of his lodgings, run to the cliff edge nearby and jumped, falling twenty feet onto the beach. There reports differ, some saying he was not seriously hurt, others that he was badly injured, lingered for a while, then died. The following year there was an auction of his belongings in Brighton but no suggestion that it was a posthumous event; rather, apparently it was more to do with Cope's finances and the notion that he was 'currently in an unhappy state at St Luke's Hospital, London'.

If, as was assumed, he was from a prosperous background, it could be that his eccentricities had led to his source of funding – a rich family from which he had become effectively estranged, perhaps – being cut off, not least because the peculiar diet upon which he insisted was costing a guinea a day, let alone the expense of his tailor-made green clothes and the specially commissioned furniture and decoration. A fund was established by 'some reputable gentlemen' in Brighton in order to help alleviate his circumstances, but the subsequent auction suggests that it hadn't been enough.

An 1808 newspaper snippet hinted that he had died in a lunatic asylum at, appropriately, Bethnal Green, and a Henry Cope was buried at St Matthew's churchyard there on 6 August 1808. There was a Henry Cope born in Bethnal Green in 1778 too – if this was the same person, he'd died at barely thirty years old.

As well as that sense of performance that's in the Brighton air, the life of Henry Cope demonstrated another of the town's thrumming undercurrents: melancholy. It's something you sense on the seafront more than anywhere else, a feeling that grows stronger the closer you come to the Channel, and I could feel it as I walked along the front between the piers. There's always something a little morose about a seaside town in winter – and here we were in the depths of November – but Brighton's melancholy is more a suffusion than a mood, just out of reach like when you try to remember a dream and it vanishes even quicker into the ether.

Is it a Channel thing? In most places I'd visited there was that tinge of sadness in the air, but it's most discernible in Brighton. The Channel being a busier stretch of water than most, it has by extension witnessed more tragedy than most, but there's more to it than that. Maybe the sad, the restless and the outcast are somehow naturally drawn to the Channel. On no other coast do you feel such a combined sense of proximity and distance: when looking out from the Channel shore, there's a consciousness of that vast landmass just over the horizon – you can even see it at the far eastern end, making the Channel a gateway to the rest of the world where, once you've crossed it, you can in theory travel as far as the Kamchatka Peninsula, the southern tip of Malaysia or Cape Town without having to cross another stretch of sea.

The Channel is the last barrier, not insurmountable but definitive, lending us isolation and possibility in equal measure. Is it coincidence that of the sixty-one surviving British piers listed by the National Piers Society, twenty-three stretch out into the Channel? The temperate climate of the south has a great deal to do with it, of course, but it's almost another expression of how we're drawn to this special 'streak of silver sea' that we have built structures to go just that little bit further out into it.

One of the most melancholic of all Channel sights is the wreck of the old West Pier at Brighton. Its skeletal, rusty remnants, cut off from the land and quietly flaking away, are home these days to nothing more than a few birds and ghosts of pleasures past. There's still a dignity in its nudity, the lines and arches of its framework retaining a noble grandeur even though we're effectively seeing the pier without its clothes. The vibrancy of the Palace Pier a few hundred yards to the east only adds to the air of sadness.

The West was the third pier designed by the legendary Victorian seaside architect and engineer Eugenius Birch, and unquestionably his masterpiece. Work commenced in 1863 and was completed in 1866, a structure 1,115 feet long and with six small ornamental buildings at the sea end serving as lounges and kiosks. Iron seating around the edge of the pier could accommodate 3,000 people. A 1,400-seat theatre was added in 1893 designed by Birch's nephew Peregrine, and an additional concert hall opened in 1916. The pier was a huge success, attracting two million visitors a year in the aftermath of the First World War. The West Pier was always the one favoured by locals, considering itself a little more highbrow than the gaudier attractions of the Palace Pier, which was the haunt of the day tripper and holidaymaker. The West was more about promenading, concerts and recitals than about palm-readers and What the Butler Saw.

The boom years of the Twenties ended with the rise of the motor car, which saw holidaymakers widening their range of destinations. Competing attractions within Brighton itself also affected pier business, especially the growing popularity of cinema. The slightly po-faced classical concerts on the West Pier gave way to lighter music, dance bands and novelty acts, but as people found more ways to entertain themselves that didn't involve sitting in front of a group of musicians the West Pier increasingly saw its plush seats lacking bums.

The rest of the pier's story consisted of game attempts to cater for users by second-guessing what they wanted. After the Second World War, during which part of the pier was demolished to prevent its use as a landing stage for the Germans, the theatre became a combined restaurant and amusement arcade and the concert hall a tea room. In the mid Sixties a local hotel bought the pier but the costs of maintenance and upkeep required just to keep the thing standing had rocketed, and by 1975 the pier had closed. After its Grade I listing in 1982 a newly formed West Pier Trust bought the structure from the council for a nominal amount, but before they could make any significant advances the great storm that swept the south-eastern portion of Britain in 1987 caused enough damage to have the pier closed off altogether to non-essential access.

This side of the millennium has seen one disaster after another befall the West Pier: another storm in 2002 washed away a section between the two buildings shortly after the trust had spent £1.5 million just on preparations for a major restoration programme, due to begin in 2003. Before anything further could be done, the spring of 2003 saw two serious fires all but destroy what remained, with another storm the following year toppling much of the structure into the sea, leaving it in the state of advanced dilapidation that now greets seafront visitors. It's a favourite of photographers and starlings – come at the right time, and you'll see some stunning murmurations – but everything about the West Pier today is at a distance, physical and in memory. The numbers of people who even remember walking on it is diminishing year on year, leaving scenes filmed on the pier in Richard Attenborough's *Oh! What a Lovely War* and 1973's *Carry On Girls* as pretty much the only way to relive the experience of Brighton's West Pier. Now, I'm a *Carry On* fan as much as the next person, but the pier's legacy deserves a little better than Sid James in a trilby and sports jacket driving along it in a go-cart.

One name that was always associated with the pier was The Great Omani, or to give him the name he was born with, Ron Cunningham. Born in Windsor in 1915 and the son of a wine merchant, he attended Sherborne public school and was set for a life of privilege in which he'd take over the family business. However, on his father's death in the 1930s the business failed, and when Ron was turned down for Army service due to a heart condition he found himself facing an uncertain future. Passing the time one day in a secondhand bookshop in London's Charing Cross Road, he picked from the shelf a volume about Harry Houdini that had caught his eye, began reading among the stacks and found his life's course newly set – he would become an escapologist. The natural showman's flair this incident released even extended to his recounting in later years that the book had fallen of its own accord from a top shelf and landed on his foot. As Ron told it, a 'paranormal happening'.

Ron lived a stone's throw from the West Pier on Norfolk Square, and from the early 1950s, once he had practised a repertoire of Houdini stunts – some in the shallow end of the local swimming baths – he became a popular draw on the pier and eventually across the whole of the south of England under his stage name The Great Omani. His repertoire was a classic selection of beds of nails, bare feet and broken glass, and plenty of fire. He would plunge into the sea from the pier hooded and fastened in chains just as Houdini had, surfacing after enough time to have passed that spectators were just beginning to worry. He would also dive off the pier into a patch of burning oil on the surface of the sea.

'People will always flock to see anybody who's likely to kill themselves,' he said of his appeal.

To mark Queen Elizabeth II's Silver Jubilee in 1977 The Great Omani performed a headstand on the cliff edge at Beachy Head with a Union flag between his toes. Age did not weary him either, and despite the decline in venues for his kind of seaside entertainment he was still performing regularly well into his eighties. Then came a farewell retirement performance on 15 July 2005 – his ninetieth birthday – at the Bedford Arms, on a street leading off the seafront in the shadow of Embassy Court.

He may have been in a wheelchair and hooked up to a kidney dialysis machine but, surrounded by glamorous assistants, he went through most of his old repertoire short of throwing himself into the Channel in chains. He crushed broken glass beneath his bare feet, smashed a pint glass against his neck with a hammer, and for the grand finale had himself bound in chains, drizzled with lighter fuel and then set on fire. That last trick was a great success until his hat went up, and by the time he realised he'd suffered burns to the tops of his ears. Pro to the last, he made it all part of the show.

'It has been a happy and colourful life, and an interesting life,' The Great Omani reflected shortly before his death in 2007. 'Otherwise, things could have been quite dull.'

The next day I called in at the Bedford Arms to raise a glass to the memory of The Great Omani and to all the acts and entertainers who'd performed over the Channel waters on the old West Pier. The Bedford Arms is all nooks and crannies, dark with drapes like a Moroccan souk. I took my pint and sat in a small room at the back and thought about people like Captain Camp, in 1860 one of the founders of Brighton Swimming Club, an organisation that did much to popularise swimming as a healthy pastime and whose members effectively used the pier as their base from its earliest days. They drew large crowds to their races and galas in the sea, people lining the pier to watch the swimmers compete and to witness feats of trick swimming and endurance.

On a wooden raft near the pier half-submerged tea parties would take place at which one of the club members would perform tunes on his concertina and read a newspaper while lying in the water, but the star turn was undoubtedly 'Captain' John Camp, who didn't let the fact he only had one leg interfere with him becoming a renowned swimmer and performer. In a way he'd lost the same leg twice – which seems careless – while working as a merchant seaman, first injuring himself and losing the leg below the knee, then suffering another accident that necessitated a further amputation at the top of the thigh. In the sea he would perform tricks such as preparing and eating a cooked breakfast of coffee, bacon and eggs, as well as feats of breath-holding and trick swimming, and all despite being limb-deficient.

Then there was Albert Huggins Heppell, stage name Professor Cyril, who worked the West Pier for six years with his curious bicycle-diving act in which he'd hurtle down a slope set up on the roof of the pavilion and fly out over the heads of the crowd, sending himself and his bike plunging into the waters below. One day in May 1912 a packed crowd watched him begin his descent, only for a wheel to wobble halfway down the ramp causing him to topple thirty feet and land with a thud on the pier decking, fatally fracturing his skull. At the inquest the coroner had branded the stunt 'dangerous and senseless, pandering to the morbid taste for the sensational', summing up just how much the West Pier had changed since its earliest days.

Less sensational but no less a draw in the late nineteenth century was Louie Webb, who performed feats in which she spent long periods under water in a glass tank, passing the time drinking milk and eating sponge cake and writing messages on a piece of slate. These curious acts stood out from the countless actors and musicians, the pierrots and strongmen, the comics and double acts, all part of a tradition that, in Brighton at any rate, died with The Great Omani. A whiskery scruff raising a pint of bitter to them in a backstreet boozer might not be the kind of tribute they all deserve, but it was the best I could do. I walked down to the seafront and crunched along the shingle towards where the pier used to stretch from the promenade out into the Channel. There had been a few iron pilings left in situ on the beach, but they were moved when the British Airways observation tower, the i360, was built, and now form a circle nearby as a kind of 'pier henge'. Maybe at the height of the summer season the sun lines up between the poles and the centre of the ruin out in the Channel, and passers-by can hear again, just for a moment, the sound of a street organ, the laughter of children and the wooden-metallic thunder of their running footsteps along the ghost of the old pier.

To pull myself out of this end-of-the-pier melancholy I made my way to Brighton's surviving one, the Palace Pier, a roaring success in the way it's embraced the classic seaside experience with its candy floss and seafood stalls, shooting galleries and ghost train. The sea boomed away beneath the planks and the ironwork, the same kind of waves that had sent me tumbling across the shingle that morning, but the sun had come out and the pier was busy with promenaders, from middle-aged couples in matching cagoules with cheeks pink from the cold to teenagers on awkward first dates, as well as those blowing away the boozy excesses of the previous night.The breeze carried the weary voices of parents calling their children back as if they didn't realise that when you're a kid on a pier you have to go to the very end, and get there as fast as possible.

The amusement arcades represent a timeless experience, the

hubbub of chatter, jingling electronic tunes, the clatter of air hockey and the cascade of coppers from the change machine destined for the penny falls, where fingertips tanned from repeated plunging into a waxy plastic pot of two-pence pieces hover, ready to launch the next coin because the timing has to be right if that teetering pile is to go over and take the little troll keyring down with it.

People watching is particularly satisfying on a pier. Everyone feels like they're somewhere familiar, yet at the same time there's something distinctly odd about walking along a giant iron-and-wood platform that doesn't lead anywhere and has the surging waters of the English Channel below. Also, everyone is happy. There are plenty of places to stop, eat, drink, browse, go on a fairground ride, but there is a compulsion to carry on walking that you don't feel back on terra firma. Is it the motion of the sea below keeping us moving? I decided to sit down and keep still for a while and think about it.

First I buy a small bag of doughnuts from a booth in which a smiling young man from Eastern Europe proudly sets the process in motion, and I can watch the action as the doughnuts pass along a machine that has 'Donut Robot' engraved on its metalwork. The little rings are sugary and hot and I take them to a table nearby. A group of middle-aged women in matching orange fleeces distribute themselves around the next table and begin talking animatedly. A young man in wraparound sunglasses wearing an immaculate white tracksuit stands by the iron railings over the sea, his arm extended, rotating his head to get the best angle for a selfie with the coastline in the background.

A bald man walks past in slacks and a sweater with sunglasses pushed up on the top of his head, accompanied by his son, in his twenties, exactly the same face as his father but sharper by three decades, a large cuddly Minion he's just won at a shooting gallery nestled in the crook of his arm – a bit of father–son bonding perhaps, re-enacting a routine from a time when both were much younger.

A small boy runs past me, stops suddenly and looks down between the planks, as if he's just realised the sea is boiling away below us. Everyone has to go around him until his parents catch up, by which time he is lying full-length on his front, spellbound by the movements of the sea below, hovering above the English Channel watching the waves that once washed around Martha Gunn in the days that made Brighton.

7

Dunkirk

The car was swaying in the wind. This was a proper buffeting, booming around the vehicle and rocking it on its wheels, all clicks and squeaks and an occasional spatter of raindrops hurled at the windows with such force they sounded like gravel, while a fine sand was starting to collect on the windscreen wipers and the runnels at the bottom of the windows. I was on the threshold of the beach at Zuydcoote, about six miles east of Dunkirk and close to the Belgian border. It was early on a January Sunday morning and mine was the only vehicle in a car park designed to hold the hundreds of vehicles that would arrive here when summer came, months hence. One thing was for sure: I wouldn't be swimming today. Not that swimming here would have felt appropriate, anyway.

Opening the car door was a challenge; it was wrenched from my hand and flung open by the wind, bouncing at the full extent of its hinge as swirling air rampaged around the inside of the car, pulling up months' worth of old parking permits and crisp packets. I heaved myself out of the driver's seat, got behind the door and shoved it closed again.

I walked down some wooden steps banked with sand and emerged onto one of the widest, flattest beaches I'd ever seen, a yellowy-grey expanse fringed in the far distance by a low and persistent parade of dark-grey waves. The wind barrelled in from the west, giving the surface of the beach a constantly shifting appearance, banners of loose sand blown across the surface like a layer of billowing gossamer. In the distance I could just make out what I'd come to see, a line of low and jagged dark shapes at the waterline that from this distance could have been seaweed or a flock of resting birds. It was low tide on a spring tide, but already the water was starting to wash around the shapes. I began to hurry towards them before they disappeared and I missed my chance.

The *Crested Eagle* used to depart from the pier at London Bridge and make her way down the Thames towards the estuary, moving southward for Ramsgate or Margate or keeping to the north side for Southend, decks crowded with day-trippers in boaters and with ribbons in their hair, waving at people on the riverside as the paddle wheels turned and the steam from the funnel hung in its wake. Launched in 1925 as the first paddle steamer in Europe to be fitted with oil-burning boilers, she was the pride of the General Steam Navigation Company, capable of carrying 1,700 passengers and with an extensive promenade deck, a sumptuous dining room on the upper deck, a smaller grill room, and a saloon at the aft dispensing teas and ice creams. Her mast was hinged and her funnel telescopic to allow her to pass under the bridges of the Thames. Then in March 1940 the *Crested Eagle* was requisitioned by the Admiralty and put into service as an auxiliary anti-aircraft coastal vessel.

In May 1940 Allied forces fighting the Battle of France had been forced back towards the Channel coast, having little answer to the German Panzer divisions that raced across the flat terrain of Flanders. Some 400,000 retreating troops found themselves pinned down in and around Dunkirk with the Germans closing in. Then the German advance stopped: so confident were they of rounding up the Allied troops, the Panzer commander ordered his divisions to pause for repairs after their thunderous journey across Belgium and northern

France. Those few days gave the British an unexpected opportunity to organise themselves and for a rescue flotilla to begin evacuating the troops from Dunkirk harbour and the beaches to its east. Operation Dynamo began on 26 May and continued until 4 June, succeeding in evacuating nearly 340,000 troops in a flotilla of 900 ships, from destroyers to small fishing boats, despite unrelenting attention from the Luftwaffe.

Bill Ridgewell was from Brentwood in Essex and was nineteen years old in May 1940. He'd left school at fourteen and found work as a bricklayer, but his heart had always been set on joining the Navy. His father had other ideas: having been wounded and gassed during the First World War, he didn't want his son potentially exposed to the horrors he'd seen. When Bill was seventeen, however, he told his father he was old enough to make his own decisions and in the autumn of 1938 joined the Navy, training for nine months on HMS *Pembroke* at Chatham.

On completing his training in the summer of 1939, Bill was posted to the destroyer HMS *Grenade*, joining the ship at Port Said and spending the summer cruising in the Mediterranean. In August the *Grenade* was sent to the Red Sea in preparation for the impending war, spending her early weeks patrolling the Indian Ocean and protecting oil tankers coming from the Gulf. In November the ship returned to Britain and undertook convoy duties out of the port at Harwich.

The *Grenade* was a happy ship, he would recall, with a popular captain, good officers, and a complete absence of bullying or ill-feeling among the crew.

Just before Christmas the *Grenade* was sent to Plymouth from where she patrolled the Irish Sea with orders to keep a lookout for the *Scharnhorst*, a German battleship that the previous November had sunk HMS *Rawalpindi* off the Faroe Islands with the loss of 239 lives. On Christmas Day the crew were just finishing their tinned pears and cream when they were called to action stations for a possible sighting. The weather had become atrocious – a Force 10 gale and waves up to sixty feet high – when the bridge saw what appeared to be a ship signalling. Thinking it was probably the *Scharnhorst* they signalled back, ordering the ship to identify itself. As the *Grenade* heaved in the heavy sea an inconclusive signal came back, interpreted by the *Grenade*'s captain as suspicious evasiveness. He ordered two torpedoes to be fired, but just before the missiles were launched someone realised they were about to attack an Irish lighthouse.

In the spring of 1940 the *Grenade* was sent up to Scapa Flow and then on to patrol duties in the Arctic Circle off Norway, where she took part in the first Battle of Narvik, returning to Harwich in May to take part in Operation Dynamo. On 26 May the *Grenade* left Dover for her first crossing to Dunkirk. They could see the smoke from the fires a long way off in the Channel, but on arrival the ship went alongside the jetty, took on a full complement of troops and managed to return without being attacked because the smoke was so dense. 'A blessing to us in disguise,' Bill called it.

Stanley Allen was a south Londoner who'd volunteered for the Navy the day after war was declared, the twenty year old enlisting on 4 September 1939, and after completing his basic training in May 1940 he and ten other new recruits were instructed to report to HMS *Windsor* at Dover. When they found the ship they learned they were replacements for men killed and wounded in the air attacks the *Windsor* had suffered while evacuating troops from the Battle of Boulogne the previous week. He arrived at the port on the 26th, the first day of the evacuation, and waited to board as the wounded and some of the dead were being disembarked, 'some two high on the upper and lower parts of the stretcher, covered over with blankets,' he recalled. When they found their mess deck the new recruits had nowhere to stow their kit, as the kit belonging to the dead and wounded was still there. They couldn't even get a cup of tea because only about half a dozen cups had survived the mess deck being hit by enemy action.

Early the next morning the *Windsor* crossed to Dunkirk. As they sailed out of the harbour and through the boom across the entrance to Dover harbour, the ship's tannoy crackled into life with Bing Crosby singing 'The East Side of Heaven', the only gramophone record that hadn't been smashed in the Boulogne evacuation.

'Every time we went from Dover to Dunkirk that same record played,' Stanley recalled.

The *Windsor* was one of three destroyers making that particular crossing, steaming gracefully across the Channel at the heart of a curious flotilla featuring a Thames tug with three empty barges astern, inshore fishing boats and a range of other small craft. The most remarkable thing were the conditions: the sea was as calm as anyone could remember, making it 'impossible to believe the Channel could be so like a millpond'. The sun was shining, it was a beautiful day, and from the gun at which Stanley was stationed he could see clearly what the *Windsor* was heading for. They'd been sailing towards Cap Gris-Nez but when they turned east, he saw a dark haze on the horizon.

'It was Dunkirk, burning.'

As soon as the *Windsor* approached Dunkirk the Luftwaffe arrived. Captain Peter Pelly, in command of the *Windsor*, managed to manoeuvre in such a way as to avoid being hit as well as protecting some of the small ships from machine-gun fire. Once the attack had finished, the little ships made for Dunkirk and the *Windsor* received orders to head for De Panne, east of Dunkirk and over the Belgian border, where they began picking up men from the beach.

Among the soldiers, Stanley recalled, was a seventeen-year-old boy wearing only a pair of soaking wet blue serge trousers.

'He turned out to be an inshore fisherman from Hastings who'd been sunk twice already that day, once on his own boat, then he'd got into another boat and that got sunk too,' he said. The crew made a bit of a fuss of him and gave him some dry clothes and boots as the *Windsor* made for the east mole, a spit of stone designed as a breakwater to protect the harbour but, with the carnage on the beach, now being pressed into service as a makeshift pier. As they came alongside, an officer with a megaphone called for volunteers to crew a fishing boat, most of whose crew had been machine-gunned. The seventeen-year-old boy who'd already been sunk twice leapt up immediately.

'There was one old seaman of thirty years' experience and he said to me, "With yougsters like that, how can we fucking lose?"' said Stanley.

After returning to Dover with a boatload of soldiers the following morning, the *Windsor* hove to off the South Foreland buoy waiting to escort some small craft from Deal and Folkestone. When the tiny fleet had assembled they set off again, accompanied by the strains of Bing Crosby singing 'The East Side of Heaven'.

Stanley and other seamen not employed on the guns set about scrubbing the decks, throwing discarded bloodstained blankets overboard and making endless piles of corned beef sandwiches. 'They wouldn't have passed in a Lyons Corner House but they'd fill a stomach,' he shrugged.

The *Windsor* picked up more men from De Panne and Dunkirk despite the attentions of the Luftwaffe, and the ship seemed to be leading a charmed life. Three destroyers that had gone over the same day had been lost and six more badly damaged. Stanley and his colleagues watched helplessly as the destroyer HMS *Wakeful* was hit by two torpedoes from a

German E-Boat, one of which hit the boiler room causing a massive explosion that tore the ship in half and sank it immediately. Of the 640 troops on board only two survived, along with 25 members of the crew.

After a night's rest, during which the crew took delivery of a consignment of Horlicks tablets as the ship had been adopted by the Borough of Slough and Windsor where the Horlicks company was based, the *Windsor* set off again for Dunkirk with Bing Crosby crooning through the tannoy speakers, heading much further east this time because German tanks had arrived at De Panne and were firing on British ships from the shore. It was a tricky route, known as 'Route Y', and involved negotiating channels between sandbanks at angles that made the ship an inviting target. Again the *Windsor* avoided damage and, finding few troops at De Panne, made for Dunkirk, where most of the remaining soldiers had gathered as the Germans began to close in.

At Dunkirk the *Windsor* took on more Allied troops than ever. Troops were forced to sit on torpedo tubes and were so tightly packed around the guns there was a genuine concern about the safety of using them. The soldiers had to help the crew stow the ropes, passing them along in a human chain, and then with space being so tight had to stand on them for the duration of the crossing.

In the meantime the *Grenade* was also at the jetty, having set out from Dover at first light and arriving around 6 a.m. They had expected a fast turnaround with a full complement of soldiers, but nobody appeared. Then word filtered back to the ship that they were being held for Viscount Gort, the Commander-in-Chief of the British Expeditionary Force and the man who had ordered the retreat in contravention of orders from London, to transport him and his entourage back to England. The *Grenade* waited nearly eight hours at the jetty before word came through that Gort was staying put, and at around 2 p.m. she finally began loading troops.

By that time, however, the wind had turned and started blowing the smoke away from the harbour, meaning Bill Ridgewell and his colleagues could see, but also be seen. Once the smoke had cleared, he noted that 'the place looked a complete shambles'.

It wasn't long before the German planes arrived and began bombing the beach – huge plumes of sand being sent into the air and drifting with the breeze, lines of men scattering, some of them dropping, others disappearing in the eruptions of sand. Still the caterpillar of soldiers loaded onto the *Windsor*, and the Luftwaffe eventually turned their attention to the other ships. Bill Ridgewell was manning a Lewis gun on the bridge and suddenly found himself flying through the air. The sound of an explosion followed close behind: the *Grenade* had come under fire again. Bill was blown upwards from his gun emplacement and hit a signaller's sponson on his way back down, injuring his back in the process.

'I couldn't walk, my back was killing me,' he recalled.

The *Grenade* had taken two direct hits from bombs dropped by a Heinkel, killing fourteen men, and immediately caught fire. Despite the pain Bill hauled himself back to his gun but found it too damaged to operate. Then he heard the call to abandon ship, managed to crawl to the main deck, and heaved himself along the gangway and onto the jetty as the flames engulfing the *Grenade* roared ever closer. Despite seeing an officer on the jetty waving a revolver around and trying to order the men back onto the ship, Bill saw another vessel coming alongside. It was the *Crested Eagle*.

The *Crested Eagle* was by that time the only seaworthy vessel in the harbour, and even that was more by luck than anything else: she'd moored on the seaward side of the eastern jetty with a rope from the bow and one from the paddle box, meaning her stern stuck out. Two bombs landed in the water between the vessel and the jetty: if the *Crested Eagle* had been moored fully side-on she would have been sunk. The Germans possibly thought they'd successfully disabled every ship in the harbour as the attack from the air ended and the *Crested Eagle* set about preparing to depart. She cast off at 6 p.m., carrying something approaching 600 troops, and with the tide falling plotted a course east along the coast among the sandbanks.

As well as men from the *Grenade* the paddle steamer had taken on board the survivors from the *Fenella*, an Isle of Man packet that had moored directly astern of the *Crested Eagle* and been bombed and sunk. She hadn't travelled far when she was attacked from the air.

'We're just getting under way and bang, a bomb comes through from the upper deck and blows up in the boiler room,' said Bill. The bomb had fallen just behind the wheelhouse, followed by two more that struck the ship towards the stern, setting the wooden structure alight and killing dozens of soldiers who had been packed onto the deck. 'There was one big flash of flame that burnt every exposed part of me,' he said. His uniform saved him from life-changing injury but his arms, face and ears were burned and he also received burns to a part of his back exposed by a tear in his uniform sustained during the attack on the *Grenade*.

For the second time in a matter of minutes Bill heard the order to abandon ship, and when he made it to the upper deck he saw the back half of the *Crested Eagle* already well alight. He scrambled onto a paddle box just rear of amidships, jumped into the water and – preferring to trust that he'd be picked up by another vessel than take his chances on the beach – began swimming for the horizon.

Lieutenant-Commander Booth, the captain of the mortally

wounded *Crested Eagle*, turned the ship's prow towards the shore and drove her onto the beach as hard as he could, so hard that she fractured her prow. That way, he reasoned, the soldiers still on board at least had a chance if they could make it ashore. But the Luftwaffe rounded on the stricken paddle steamer and strafed the survivors. Of the estimated 600 on board, half were killed.

Back on the *Windsor*, Stanley Allen could see what was going on and didn't like what he saw. He watched the men struggling on to the beach, many of them already badly wounded, only to be met with a hail of machine gun bullets from the air.

'It wasn't cricket,' he said. 'That wasn't the right way to win a war, having a go at wounded people.'

The *Windsor* was one of the last ships to leave at the end of the evacuation. It was getting dark and just as they began to clear the harbour of Dunkirk the wind blew up and the sea began to swell dramatically.

'It was remarkable,' said Stanley, 'it had been so smooth and suddenly on what turned out to be the last day it started getting rough. It was like a miracle.'

Back at Dover, the *Windsor* took on oil and water as usual, moored at a buoy, and then heard Operation Dynamo was officially over. They wouldn't be going back, a realisation that had a curious physical effect on the men.

'Old sailors, and me, everyone, we just started to shake,' he said. 'It was a post-battle reaction.'

Whenever he was asked about his memories of the Dunkirk evacuation, Stanley always spoke of this sheer exhaustion he and his colleagues felt in the immediate aftermath. He was too tired to do anything, could barely summon the energy to eat and even found sleeping difficult. That was his overriding impression, along with the fact crews didn't panic under pressure because they were just too busy to think about anything but the immediate task at hand.

'You just accepted that if your name was on something you were going to buy it,' he'd shrug.

As for the wider implications of the evacuation, Stanley remembered thinking about all the men, equipment and ships that never came back and experiencing a chilling vision of the plausible end of the only way of life he knew. It's hard to appreciate today just how close Germany came to winning the war in what turned out to be its early stages. The Dunkirk evacuation proved crucial in the eventual Allied victory, but at the time winning the war seemed a distant prospect.

'We just didn't know if we could hold Jerry off,' recalled Stanley of that summer in 1940.

Bill Ridgewell meanwhile, suffering from burns and his back injury but still swimming, was eventually picked up by a motor boat that transferred him to a minesweeper. He was taken by stretcher to the captain's cabin, where he was given a tot of rum before a medical orderly arrived with strips of lint soaked in cold tea which were placed on his burns. The next thing he knew he was at Ramsgate, then a hospital in Dartford, before being transferred to a psychiatric hospital in Belfast where it was recommended that he never go to sea again. Despite his ordeal and the opinions of his doctors, almost exactly a year after Dunkirk Bill was back in the Navy where he served for the rest of the war.

'I've no regrets about serving in the Navy,' he said. 'If I'd still been fit when the war ended I'd have stayed on.'

I was still striding over Zuydcoote beach, hoping to beat the tide before it covered what little remained of the *Crested Eagle*. It's visible only rarely, on spring tides at low tide, and even then the Channel only allows it to be visible for a matter of minutes. I was slightly later than the lowest point of the tide and as I hurried over the sand, its wraith-like patterns still shifting over the surface making the wind visible, I was concerned I wouldn't make it. Splashing through puddles of clear water over compacted ribbed sand, I made it just in time. Jagged rows of broken iron teeth are all that protrude these days, all that remain of the paddle steamer that gouged its way onto the beach in flames as Luftwaffe bullets made lines of spume on the water. After she'd run aground her fuel tanks exploded, making any hope of survival for men still on board impossible. People on ships nearby that tried to get close were driven back by the fearsome heat.

The beach where I stood would have been packed with clusters of soldiers awaiting rescue, barbed wire, the odd tank trap, the debris of war and rescue: metal cases, bomb craters in the sand, the thump of German shells and the scream of German aircraft. That morning I was the only person on the beach and the only thing visible on it was the iron outline of the *Crested Eagle*. In the middle of that night in the spring of 1940 the heat from the fire had been so great that even when it had all but burned itself out the hull glowed orange in the dark.

A plaque has been affixed to the prow, unveiled at a memorial service that took place in 2015 to mark seventy-five years since the Dunkirk evacuation. The ceremony was attended by ninety-eight-year-old Victor Viner, who had been on the beach trying to help coordinate rescue efforts as the *Crested Eagle* burned, unaware that his brother Bert had died on board, having transferred onto the paddle steamer from the *Grenade*.

The wind kept barrelling along the beach from the direction of Dunkirk, roaring in my ears as the incoming tide began to lap around my shoes. I looked along the hull of the *Crested Eagle*. She's buried up to the waterline, a phantom of a ship, an imprint of a tragedy – meaning that the remains of some of the men who'd boarded her hoping to be taken across the Channel to safety that spring day are probably still somewhere beneath the sand. Above the sound of the wind I became aware of a flurry behind me: a flock of birds about twenty strong that flew over my head and alighted on the wreck, wrapping their feet around the ironwork, tucking in their wings and looking haughtily at the shallow waves washing in around the rusting metal.

I turned into the wind and walked back across the sand, climbed the wooden steps and went back to the car. Slamming the door from roar into silence, there was a blasted rawness in my eyes, the sand grains that had forced themselves between my teeth crunching loudly inside my head.

It's hard to think about Dunkirk without thinking about, you know, *Dunkirk*, but that astounding averting of complete disaster that could quite possibly have handed the war to the Germans has become such a key part of the British narrative that it's hard to separate town from evacuation. So instilled has the 'nine days' wonder' become in our national culture, a retreat with heavy losses against odds that made it feel more like a victory, that it's easy to forget that there is also a town with a history behind the name.

Indeed, Dunkirk – the northernmost Francophone town in the world, fact fans – has been a busy place over the centuries, not least in trying to accommodate various different countries looking to own it. It's a strategically important spot: Dunkirk has over the centuries been shuffled around the Dutch, the English, the French and the Spanish.

For six years during the 1570s and 1580s the place was in the control of a group called the Dunkirkers, a bunch of Spanish-backed Catholic privateers with more than a hundred ships. These ships were crewed by sailors so skilled that when Dutch Protestants blockaded the port, the Dunkirkers just snuck or battered their way past or through the blockading ships and set about relieving the busy trade traffic passing through the

Channel of their cash, other valuables and cargoes. Sometimes they'd even sail as far as the Baltic and the Mediterranean in pursuit of swag. After centuries of squabbling between the various Dutch and Spanish states, in 1658 Dunkirk was captured from the Spanish after a month-long siege by an Anglo-French force under French command, and immediately handed over to the British. On 25 June the Spanish forces opened the gates and left, handed the city keys to the French, keys which Louis XIV personally passed on to the English. Thus, on that day in 1658 Dunkirk was under the sovereignty of three different nations.

The problem was, though, that after years of civil war England was not exactly rolling in cash, and even less so when Charles II was restored to the throne in 1660. Dunkirk, garrisoned by an uneasy mix of former soldiers of the New Model Army and royalist battalions from Ireland who had been loyal to Charles all through his exile, was a luxury England couldn't really afford. And anyway, in the meantime they'd managed to secure the port of Tangier, which was infinitely more exciting and exotic than boring old Dunkirk.

Another issue that had Charles and his advisers chewing a pensive lip was the fact that possessing territory on the other side of the Channel meant there was a fair-to-middling chance of being drawn into the many wars that rampaged back and forth across the continent, conflicts that the Channel permitted England merely to watch from the international-relations equivalent of an executive box. Charles had no qualms in putting Dunkirk up for sale and, having flicked through the pictures on Rightmove, in October 1662 Louis XIV forked over five million *livres* for the town that Charles probably immediately spent on wigs. It wasn't a popular move (despite the glamorous distraction of Tangier), Samuel Pepys noting in his diary that he was 'sorry to hear that the news of the selling of Dunkirk is taken so generally ill, as I find it is among the merchants'.

Dunkirk is not a pretty town today by any stretch, but that's understandable given that it was almost completely flattened during the Second World War. There is a slightly weary pride about the place, though – a sense that it wants to sit you down and explain that it may not look like much, but wait until you hear why. War and fortification have been the running themes throughout the history of the town: Louis XIV set about building fortifications, extra forts, expanding the harbour and improving river and canal access almost as soon as Charles's cheque had cleared, but most of these improvements were destroyed, not by a besieging army but by paperwork: the Treaty of Utrecht, drawn up in 1763 after Britain defeated France and Spain in the Seven Years War, specified that Dunkirk must be rendered pretty much defenceless, as the British felt that if anywhere was going to be used at some point to launch a cross-Channel invasion, then a heavily fortified town within easy sailing distance might be the place to do it.

Dunkirk suffered during the First World War too. As a key port in transporting troops and supplies from England, from 1915 the town came under regular bombardment by the Germans' long-range guns, and in 1917 the town was well within range of the biggest gun in the world, the Lange Max, a fifteen-inch behemoth firing shells that weighed nearly a ton, constructed at Koekelare in Belgium and firing on the town from more than thirty miles away. On clear nights the flashes of the exploding megashells could be seen from Dover. Add to that bombing raids from the air and shelling from the sea by the German Navy, and it's amazing that only 600 people were killed in the town. There was a population exodus, however: during the first three years of the conflict Dunkirk's fell from a shade under 40,000 to around 7,000, but the bravery of the town was recognised in 1917 with the award of a Croix de Guerre and later the Légion d'Honneur.

For locals, the Second World War isn't so much about the evacuations of 1940 as about the Allied bombardment during the German occupation that all but razed the town, followed by an eight-month siege between September 1944 and the end of the war in May 1945. After the success of Operation Overlord following the Normandy landings, the Allies pressed on eastwards through France but found the Germans embedded in a heavily fortified Dunkirk. The thing was, the Allies weren't that fussed about it: they'd taken Ostend and were opening up access to the River Scheldt so as to supply Antwerp. The port at Dunkirk was so badly damaged as to be useless anyway. As the Germans were pushed back east their garrison there became more and more isolated until there was merely a token besieging force during a tough winter for both sides. A truce was agreed in October 1944 to let the town's 17,000 inhabitants leave, meaning the German soldiers were then rattling around in a bombed-out ghost town during a cold, wet winter with the war turning against them. Beyond a few token nibbles by the besieging Czech units to keep the Germans on their toes, the Allies were happy to ride out the war around Dunkirk and on the day of the German surrender their garrison called it a day and the town was back in French hands.

Like Calais, the postwar rebuilding of Dunkirk prioritised substance over style, and much of the town is a hotchpotch of identikit modern structures. There isn't much visible evidence of war left today beyond a spray of bullet scars on the impressively no-nonsense frontage of the Catholic church of Saint-Éloi. Dunkirk could never be described as picturesque, but that's almost a selling point. The fact that it's a flourishing, lively town, given all the hardships it's endured, the worst of them still in living memory, and still has a distinct personality demonstrates a core determination rooted in the soil and the sand.

It was close to Christmas and there was a grotto to be visited in the commanding, beautifully restored town hall for which a queue of excited children stretched down the main staircase, out of the doors and along the street (at least, I assume it was a grotto they were there for, unless the children of Dunkirk have a keen interest in town council planning meetings). On the nearby Place Jean Bart was a Christmas market, which in some towns would have been a rather miserable experience on that cold, windy, rainy evening, but in Dunkirk it was an absolute joy. It might have been to do with the sheer number of small wooden cabins knocking out grog – every other establishment seemed to be offering either mulled wine or hot spiced cider, and it took a while for me to get round all of them, I can tell you. It might have been the booze, but the unremitting happiness of the occasion in conditions that were hardly ideal seemed very Dunkirk to me. It's been attacked, bombed, besieged, pillaged, fired and ransacked, yet its people have remained stoic and cheerful, displaying the same imperishable nature as the town itself. Wandering among the cabins and stalls, lips sticky with mulled wine and with breath about 80 per cent clove, I was overcome by a wave of affection for this battered old Channel town.

At the heart of the Christmas market, in the centre of the square, was one of those statues that pop up all the time in French Channel towns. Bathed from below in yellow light was another man in thigh-length boots and wide felt hat, but this one had a flamboyant plume on it, not to mention that the figure, holding aloft a short sword, stood turned in the direction of the English Channel – and with a considerable degree of élan, at that. The name on the plinth was that of Jean Bart, the epitome of Dunkirk and one of the greatest Channel people of them all.

A brilliant seaman and a privateer of considerable success, Jean Bart was born in Dunkirk in 1650 to a Spanish mother and a soldier father who died in battle fighting for the United Dutch Provinces against the English in 1668. His Channel lineage was quite something: his grandfather Coril Weus was a sea captain who fought the Dutch for the Spanish side based out of Dunkirk during the Eighty Years War. His great-grandfather Michel Jacobsen was also a privateer, and a respected pilot who shepherded what remained of the Spanish Armada back to safety in 1598. Indeed, he was such a great servant to Philip of Spain that in 1624 he was awarded the Cross of St James for his actions against the Dutch and the English, who nicknamed him 'the fox of the seas'. In 1622 his great-uncle Jan Jacobsen blew himself up in the Channel, along with his ship, rather than surrender to the Dutch.

Given such a feistily briny background it was inevitable that Jean Bart would go to sea, and he joined a corsair crew at the age of eleven in 1662, the year Dunkirk was sold to Louis XIV. By 1667 Bart was on the crew of the *Seven Provinces*, a ship of the line that took part in the audacious Dutch raid on the River Medway that scored impressive successes against British ships in Chatham and Gillingham, towing away the Navy's flagship the *Royal Charles* in the process.

When Louis XIV declared war on the Dutch in 1672 Bart joined the French side and began building his unparalleled reputation for harassing shipping in the Channel and the North Sea. Not being of noble birth, he couldn't rise to the rank of officer in the French Navy, but as captain of a Dunkirk privateer he unleashed carnage, fighting in six battles and capturing eighty-one ships – the kind of success that forced the French to change the rules, accept Bart into the French Navy and make him first a lieutenant and then, in 1686, the captain of the frigate *Railleuse*.

Two years later in the War of the League of Augsburg, France allied with Denmark and the Ottoman Empire against the combined forces of England, Germany, Spain, the Netherlands, Savoy and Sweden. When Bart's small fleet was escorting some captured Dutch merchant ships up the Channel to Dunkirk, he was engaged by the British off the Isle of Wight, wounded, captured and imprisoned in Plymouth, along with one Claude de Forbin, another wounded French officer. At first the two men were treated harshly, not least de Forbin who was stripped naked, and Bart was only allowed to remain clothed because he spoke English. The Governor of Plymouth took a shine to the pair, however, and would dine with the two Frenchmen, de Forbin still in the altogether, before placing them in an apartment above an inn that had had barred windows fitted and sentries posted outside. When a fisherman relative of Bart's from Ostend was forced to put in at Plymouth during a storm, he heard of Bart's incarceration and, on visiting him, managed to pass him a heavy-duty file. Bart set about sawing through the bars at their window as surreptitiously as he could, aided by the fact that the sentries, finding themselves posted at an inn, spent most of their watches royally hammered.

The surgeon who came in to dress their wounds was a Frenchman who had been taken prisoner at sea and was keen to go home, so he was in on the escape plan too. The Ostend relative was under instructions to look out for a suitable boat, so when he came across a rowing boat containing a sailor dead drunk and snoring like a walrus, he lifted the man out, hid the boat, ran to the inn and told Bart it was time. Careful to dispense the necessary aid to the still badly injured de Forbin, the two prisoners and the surgeon slid down knotted bed sheets, scuttled off to the harbour, found the boat and set out in the direction of France under cover of night and a perfectly timed fog. When a patrolling coastal cutter at the harbour mouth called out demanding to know what kind of boat they were, Bart's English was good enough that his answer 'Fishing boat!'

convinced them, and out they went into the Channel. For two and a half days they heaved away until finally the trio landed exhausted not far from St Malo, where they learned from sentries on the coast keeping an eye out for Huguenots trying to escape to England that everyone thought they were dead. For that adventure alone, Jean Bart would be one of the heroes of the English Channel, but there was still more to come.

In the 1690s he began setting about the Dutch herring fleets in an effort to starve the nation out, on one famous occasion battering his way through an attempted blockade of English ships designed to stop him – his seven frigates and a fireship proving superior to the greatest navy in the world. He terrorised the English and Dutch merchant fleets, captured two ships laden with expensive goods bound for Russia, and burned out one Scottish castle and four villages just because he was on a roll.

Arguably his greatest achievement, however, came in June 1694. Two successive grain crop failures and a successful blockade by France's enemies had left the country in a state of famine that caused thousands to starve to death. The French government arranged for a convoy of grain ships to sail from Norway and sent Jean Bart with nine ships to escort them to France. The 120 grain ships didn't wait for Bart, however, and were captured by the gleeful Dutch. Bart patrolled the North Sea looking for the convoy and eventually found them just off the Dutch island of Texel. He went straight for the Dutch flagship in charge of the convoy and within half an hour had outwitted, outmanoeuvred and outgunned the Dutch commander Hidde de Vries, who later died as a result of his wounds. Bart sailed the grain ships to Dunkirk where he was greeted by crowds celebrating with wild abandon: he had almost literally saved the nation, bringing home enough grain to avert the worst of the famine. Louis XIV made him a Knight of the Order of St Louis and elevated him to the nobility. Two years later he outwitted the combined nous of British and Dutch commanders at the Battle of Dogger Bank, again forcing his way successfully through a blockade to bring home twenty-five captured merchant ships and 1,200 prisoners.

In 1702, having been promoted to squadron commander after his Dogger Bank success, he was putting together a fleet ready for the War of the Spanish Succession when he fell ill with pleurisy and died at his home on 27 April. He was buried with full honours in the Church of St Éloi, in whose direction his statue is looking today.

In 1845 the town unveiled the statue and a two-day festival was held at which a specially composed song was performed, something repeated annually in Dunkirk to this day. Twenty-seven ships of the French Navy have been named after him, the current *Jean Bart* being a frigate launched in 1985.

I raised my steaming cardboard cup to Jean Bart, legend of the English Channel and the man to whom and around whom Dunkirk can always rally. The following day before I left the town I went into the tourist office and bought a tiny snow globe with 'Dunkerque' on the pedestal and a miniature Jean Bart statue inside. When the ferry back to Dover sailed out of the harbour I almost felt it necessary to apologise that we hadn't had to blast our way out through a blockade.

He's on my desk now as I write this, tiny and fierce, facing the Channel with his sword raised. If someone knocks on the door saying they're a friend of his from Ostend I'll be sure to frisk them first.

8

New Year

Eight o'clock in the morning, January the 1st, a little bit of a hangover, I started the year with a swim in the Channel. I can never sleep in after a few drinks and there had been less than ten minutes between my waking up and standing on the threshold of the Channel on the first morning of the first day of the year. My mouth was dry and tasted metallic and a headache was making its way from the back of my head round towards my temples, but my first thought on opening my eyes was about getting into the water. The conditions were absolutely perfect. The tide was high, there was no wind, the sea was a sheet of glass and, to the north, weak yellow sunlight was breaking through a tear in the thin layer of cloud. For once I had company on the shore – further along the beach a distant fisherman sat in front of his tent, wrapped up against the cold in a woolly hat and all-weather gear, a radio at his feet, rod propped on the shingle, and reaching down to pick up his flask.

I unzipped my fleecy robe and dropped it onto the stones next to my tea. My Thermos mug of tea was becoming more than a luxury on these cold mornings: the Channel was cold enough now that when I came out and swallowed a hot mouthful I could feel it travel all the way down into my stomach. I walked down the last few feet of shingle and let a tiny wave lap over my feet, the water so cold it went through me like an electric charge. I waded in further: legs are the easy bit, it's when the water gets above the waistband of my shorts that the cold really hits and my breath starts making noises like an asthmatic coffee percolator. I'd like to pretend that I glide serenely into the water, smiling beatifically, but in reality I often find myself shouting swearwords I didn't even know I knew.

As usual the sea began to feel pleasant once my shoulders were under. My breath came fast and shallow and I held it for a moment before exhaling, then pushed off from the bottom and started swimming towards the first sunrise of the year. The Channel hadn't reached its lowest temperature yet – that was still a few weeks away – but it was cold enough to burn. I could feel it immediately on my arms and torso like I was standing close to a sheet of flame, a feeling that had become a little bit addictive. I could still feel it when I'd left the water, had a hot shower and even when I'd sat down to work, a fizzing reminder on my skin of how I'd started the day.

There was a slight haze on the horizon but I could still just make out the outline of the French coast. It's been too foggy or hazy in the years I've been in Deal, but they say that when the New Year's Eve night sky is clear you can walk down to the shore away from the light pollution at 11 p.m. and see the fireworks being let off an hour into the future on the other side of the Channel.

Although I'd undoubtedly built up a resistance to the cold over the previous months, I still after only about ten minutes needed to worry about getting *too* cold. My arms thrust out in front of me and swept around: I like breaststroke best as I can see where I'm going and can take in the Channel view, eye level along the surface all the way to the horizon where the container ships pass so slowly it's like they're barely moving at all. When I face north I'm swimming towards Deal Pier and, in the distance, Ramsgate; south, it's towards the start of the White Cliffs after Kingsdown, where I watched that morning as a ferry headed out of Dover harbour while an incoming vessel prepared to replace it, our relentless connection to the European mainland not at rest even when the nation is suffering from its collective annual hangover.

I also prefer breaststroke because it's practically the only one I can do. I am a swimmer in the same way that someone who can pick out the first few notes of 'Happy Birthday' is a pianist, but in the Channel that doesn't matter. I'm not swimming to go anywhere, I'm not trying to meet any physical goal and I'm certainly not competing with anyone. I love the Channel, love living by the Channel and love being in the Channel, and that's it. No macho nonsense, no midlife crisis, just the simple daily pleasure of being in the water.

There was little tidal pull that morning, so I swam around in circles for ten minutes and made my way back ashore, feeling my limbs grow heavy again as I sloshed out of the water. Even though it was cold and I should have put on my fleecy robe as soon as possible, on mornings like that, when it's cold but almost completely still, I can't help standing for a moment to feel the air moving gently around me. On calm days like this being cold is a thrilling sensation, and swimming through that winter I found myself experiencing something that had always been alien to me: a kind of bravery. I felt brave heading out into a freezing dawn, then walking into the granite-cold English Channel until I was out of my depth and swimming. I felt brave unzipping my fleecy swimming robe and standing there in just shorts and a pair of thin rubber shoes preparing to go in. Bravery is not something I've ever really done. Indeed there is a corner of south-east London that is, to me at least, forever frightening, a corner with a direct connection to my morning dips in the Channel.

Eltham, a suburb of that part of London, has on the face of it quite a lot in its favour. It's leafy, or at least, leafy as far as south London goes. It's got good transport links, a half-decent high street, and it pops up semi-regularly these days in newspaper property supplements as the next London hotspot for craft beer entrepreneurs and fans of coffee that doesn't come in granules. Bob Hope was born in Eltham. E. Nesbit lived there. Frankie Howerd spent a chunk of his childhood there, Boy George grew up there and was expelled from the local school, and W.G. Grace spent his last years there.

I spent a lot of time in Eltham as a boy, and most of it wasn't frightening. My lovely Great-aunt Ruby lived not far from the high street and I used to walk past Kate Bush's house on the way to hers from the station. It was near enough to school that you could bunk off to McDonald's or the Our Price record shop at lunchtime and be back before anyone noticed you'd gone. It was all right, was Eltham.

Eltham was also home to Eltham Baths, and that was the frightening bit. Eltham Baths were housed in an austere-looking 1930s brick building with a square tower that made it look a bit like a church, or perhaps a concentration camp. I've just looked at a picture online and after all these years it still gave me the same cold sensation in my stomach as it did back when I was a skinny kid approaching the place gingerly with his trunks rolled up in a towel. I'm as community-minded as the next woolly liberal, but I have to admit that when I read beneath the photograph that Eltham Baths had been demolished in 2011 I allowed myself a little fist pump.

When I was a small boy I learned to swim at Eltham Baths. I was scared of most things back then – off the top of my head: ghosts, the dentist, the doctor, the cupboard under the stairs, the Incredible Hulk, the alley behind our house, horses, most of the teachers at school, wasps, motorbikes, the elderly couple that ran the local corner shop, car washes, the swings at the park down the road, dogs, cats, tortoises, thunder, trains and heights.

In the light of all that, my fear of Eltham Baths was almost rational: I couldn't swim and was being sent somewhere that was a temple to swimming in order to put that right. I don't remember much beyond the terror, not even who was teaching me nor anyone I was learning with. My memories are quite impressionistic: the stomach-churning fear as I pulled on my swimming trunks and waded through the little disinfectant footbath between the changing rooms and the pool, the cold-ness of the steel steps as I grasped them ready to get in, the battered polystyrene floats we were given to hold out in front of us while learning to kick our legs but which I just used to hug close to my chest at the side. I was not, in short, ever going to be one of life's swimmers.

The only thing I really remember with any genuine clarity is a glass-fronted noticeboard at the entrance, in front of which I'd stand shoving vending-machine crisps into my mouth with fingers still ribbed from immersion that lent a hint of chlorine to the salt and vinegar, and wait for my dad to come and pick me up. The board was taken up entirely with yellowing newspaper cuttings and black-and-white photographs of a lad in goggles, trunks and a rubber swimming cap. In one of the pictures he was standing up to his thighs in water and smeared all over with some kind of greasy substance; in another he was swimming, goggled face turned sideways, mouth agape for breath, his arm coming over the top with fingertips arrowing towards the water. The picture I remember most clearly is of him standing in his trunks and swimming cap with his arms raised in triumph, gog-gles pushed up onto his forehead, a big smile on his face. The headlines on the accompanying clippings trumpeted the news that a local lad had swum the English Channel.

A rummage in the records reveals the boy in the pictures to

be Marcus Hooper. He had indeed swum the Channel, from Dover to Sangatte on the French coast, on 5 August 1979 in a time of fourteen hours and thirty-seven minutes. That's impressive enough in itself, but it's a startling achievement when you consider that at the time Marcus Hooper was twelve years old. He was from Kidbrooke, not far from Eltham, and had done most of his training at Eltham Baths as a member of the Eltham Training and Swimming Club. Not only had he swum the Channel, he'd broken a record that had stood for all of twenty-four hours: the previous day another twelve-year-old, Kevin Anderson from South Africa, had made a successful swim in twelve hours. Marcus Hooper was two hours slower than Kevin Anderson, but he was three months younger, which grabbed Marcus the headlines before Kevin Anderson's trunks were dry.

'I am sorry for Kevin but I had to do it,' Hooper told the press the next day. 'I was stung four or five times by jellyfish on the nose, stomach and legs and I yelled out but I knew I had to go on.'

There I was, squinting up through misty NHS glasses at a kid not a million years older than I was who'd just swum across the most famous and demanding stretch of water in the world while being set about by jellyfish – of which I was also scared – and I was still too timid even to get into the kids' pool by any means other than the steps. While Marcus Hooper had just swum the Channel, a certificate for completing one width of the same pool he'd spent hours ploughing up and down preparing for his epic swim was still a pipedream for yours truly.

As it happens, that particular part of south-east London where I cut my marine teeth has a notable Channel-swimming legacy. Some might even call it a hotbed. Mostly that's down to a man called John Bullett, who ran the Eltham Training and Swimming Club in the Seventies and Eighties and also trained Tom Gregory to swim the Channel in twelve hours in 1988 at the age of eleven. After that the Channel Swimming Association, who along with the Channel Swimming and Piloting Federation are the official arbiters, imposed a minimum age of sixteen for aspirants, so that's one record that will never be broken.

It's not only Bullett's protégés who've set off for Dover from this corner of south-east London with a towel over their shoulder and a big tub of lanolin in their coat pocket, either. As far back as 1928 Hilda Sharp, aged eighteen, became the third British woman to swim the Channel and at the time was fifth fastest ever, crossing in fourteen hours and fifty-eight minutes. Known as 'Laddie' because she had short hair and a boyish face, Hilda was from Hither Green, a stone's throw from Eltham. Newsreel footage of her swim shows her in its final stages looking exhausted, her arms barely lifting out of the water as she stroked. Then she's being helped from a boat onto the steps at Dover with a heavy overcoat round her shoulders and a beret on her head. Film of her homecoming shows the street where she lived jammed with people waiting to greet her, then the area outside Lewisham Town Hall equally thronged as the mayor presented her to the crowds.

I thought about Marcus Hooper and Hilda Sharp as I zipped up my robe, twisted open the lid of my mug and sucked down a mouthful of hot tea. Dribbles of sea water ran from my hair down my face as I pulled up the fleecy hood and looked out to the horizon on the year's first morning. It struck me then that everywhere I've ever swum has either been *in* the Channel or had a connection with it. I'd shared these waters with south-east London Channel greats like Hilda Sharp, Marcus Hooper and Tom Gregory and shared the waters of Eltham swimming baths with the last two, possibly at the same time.

I couldn't really see myself swimming anywhere except the Channel. The more I swam the more I felt I belonged in it. I was comfortable with how, despite it supporting me, embracing me, giving me space from the ritual of the everyday, allowing me to forget the stresses and fears of the real world, lending me utter solitude with just the sound of my breathing and the trickle and splosh of the water around me for company – the English Channel didn't give a toss about me, my swimming or my welfare. It was an arrangement with which the Channel and I were absolutely happy.

9

The Channel Swimmers

There are mornings when I make my way across the shingle for my morning swim when it's clear enough to see the coast of France. In the evenings with the sun from the west you can make out the white cliffs themselves and sometimes even the green fields above, but at dawn it's generally a grey mass low on the horizon like the back of a giant whale.

Sometimes when I'm in the sea I'll stop and look across the Channel to that low-slung outline and think about just how close France is. Then I ask myself the same question that friends ask me whenever I tell them about my morning swims (something I do often and at length – they're very patient). Did I ever think about actually swimming the Channel? Weighing up the pros and cons, how acclimatised to cold water I've become, how much stronger a swimmer I now am than I've ever been, how dramatically my knowledge of the tides and the weather has increased since I took up sea swimming. I suck on my bottom lip for a while, then give myself the same answer I give every one of those friends.

Fuck, no.

It would be like asking someone who can just about run to the post office before it closes whether they'd consider the London Marathon. Not only that, but running it in freezing temperatures while spectators lob jellyfish, saturated planks and the odd shipping container at them. And that's before even considering that I'm a middle-aged man with the muscle tone of Charles Hawtrey and the coordination of a prop forward on ice skates.

There aren't many extreme-sports enthusiasts that don't have me tutting and rolling my eyes at their shiny-faced, sinewy smugness, but Channel swimmers are different. The people who cross the English Channel purely by the power of their own limbs are a special breed, particularly when you consider the regulations by which they have to abide for their swim to count as officially recognised by either the Channel Swimming Association or the Channel Swimming and Piloting Federation.

Swimmers are not permitted to wear wetsuits, for example. They can wear a regulation swimming costume, swimming cap and goggles, but that's it. They're not allowed to touch or be touched by anyone, nor can they at any time touch their support vessel (food and drink is usually administered via a fishing rod). They also have to walk a certain number of steps unaided once they leave the water at the other end for their swim to count. Then there's the distance to think about. Swimming twenty-one miles would be enough of a challenge in a heated pool, but the fast-moving Channel tides are washing constantly across the swimmers' paths, pushing and pulling them considerable distances from the straight and narrow. A successful crossing involves swimming in a broad S shape of forty miles or more as the tide pushes you first one way, then the other, and you spend many hours in the water: the current record for the shortest crossing is six hours and fifty-five minutes, while the longest single crossing on record took a shade under twenty-nine hours.

As if crossing once isn't enough, there have been several instances of swimmers making it a double or a triple. In 2019 an unprecedented four-way crossing was achieved by thirty-seven-year-old American Sarah Thomas, who covered an estimated 130 miles while spending over fifty-four hours in the water – and all just a few months after completing treatment for breast cancer. The regulations for Thomas's swim were the same as for everyone else: in those fifty-four hours she couldn't leave the water at either end, she had to begin her return swims within ten minutes of completing the previous one; no one could touch her, no towelling her down, no reapplication of insulating substances on the parts she couldn't reach, no massage – not even, for goodness' sake, a hug.

It takes something very special indeed to be a Channel swimmer – a rare combination of physical stamina, technique, hardiness, bravery and a strong streak of bloody-mindedness – so considering what an incredible achievement it is it remains a fairly under-recognised accomplishment. The traditional way of marking a successful crossing is a pint in the White Horse pub in Dover and, if there's still room, the writing of your name, the date and the crossing time on the pub wall. Most crossings today are made to raise money for charity, and beyond a smiley, ruddy-cheeked, knackered picture of you in your local paper there's isn't a great deal of hullabaloo, just personal satisfaction and a sum of money winging its way to a good cause. At the time of writing, the four-way swimmer Sarah Thomas has fewer than 700 followers on Twitter. There are spittle-flecked trolls arguing with themselves who have more than that.

If I were to claim that as someone who swims in the Channel every day I feel an affinity with the Channel swimmers, it would be a bit like claiming a picture of a daisy drawn in biro on an old envelope gives me an affinity with Vincent van Gogh. These people are extraordinary – I barely qualify as ordinary. What I do feel is a deep respect for and fascination with the hundreds who have made the crossing since it was first achieved as long ago as 1875. Whatever brief, spluttering immersions in the Channel shallows I might accomplish, they're not in the same galaxy as the Channel swimmers' feats.

Given that I would never be joining their ranks, then, I wanted to find out more about them. In particular I wanted to learn about the pioneers, the ones who not only swam the Channel but did it before anyone else had. Striking out from the shore and making for the grey outline on the horizon being out of the question, I began out of sight of the sea, high up on a hill and in a cemetery, where I arrived with – thanks to a kind woman in the burial-registry department of Dover District Council – detailed instructions for locating a particular grave. Having parked the car I pulled out my phone and was searching for the email she'd sent me when there was a tap at my driver's window. I looked up to see a bearded man in dark clothes, smiling at me and motioning that he wanted to say something. Flustered by the unexpected interruption, I proceeded to lower the three other windows before managing the driver's one.

'Hi there,' he said. 'Sorry to bother you but I just wanted to ask – do you mind if I fly?'

'Do I mind if you—?'

'Fly,' he confirmed, still smiling.

'Erm, no,' I said. 'Not at all. Be my guest.'

'Thanks very much,' he said. 'I thought it best to check.'

Expecting to see him begin slowly flapping his arms before ascending gracefully into the heavens, still smiling at me, instead I was slightly disappointed to watch him jab at a console to send a small drone shooting skyward with an airy whine.

It was nice of him to ask. I could have been visiting a much-missed relative for a period of quiet contemplation and might not have appreciated a plastic object swooping low over my head as I pondered the great eternity, but I wasn't. Instead I was there to see the last resting place of a man to whom I was not related and who had shuffled his way from this earthly plain 120 years earlier. It wasn't even someone who'd swum the Channel, but as landmarks from the pioneering days of the crossing go it was as significant as any I'd find today. It was a much larger grave than I was expecting, three plots wide and landscaped in green shingle, its many occupants commemorated on three white headstones with, at the centre, 'In Loving Memory of George Toms who died 21 August 1900 aged 76 years'.

Further down was another inscription.

'Pilot to Capt. M. Webb,' it said, 'who swam the Channel, 25th August 1875'.

George Toms was about as Dover as it got. He died at his home on Park Street in the town, a stone's throw from where he was born. As a boy he was soon at sea on a fishing smack and then on the colliers bringing coal down from Newcastle, a dirty job that usually entailed little more than endless working of the bilge pumps. Eventually Toms acquired his own boat, the lugger *Anne*, and began ferrying small cargoes around the coast and across to Calais as well as using his vast knowledge of the tides and vagaries of the Channel to become a sought-after pilot for ships negotiating their way in and out of Dover. So lucrative were these endeavours that he was able to secure the lease of the large Shakespeare's Head inn in Dover to supplement his maritime income.

It was no surprise that Matthew Webb sought out Toms' services to accompany him on his attempt in 1875 to become the first man to swim the Channel. Webb was a seaman himself and he chose wisely: Toms remained the voice of calm reason and certainty throughout, countering any suggestion that the swimmer might be fading. Especially in the last couple of hours, Toms used all his knowledge and experience to keep the *Anne* in the best possible position to shield the exhausted swimmer from the worst of the wind. Despite nobody ever having swum the Channel before, Toms remained certain all along that Webb would do it even when he looked to be struggling. Webb presented him with a gold ring in thanks, and Toms immediately became the first choice of pilot for all subsequent attempts on the Channel at the time. But Webb was the first, and George Toms had a lot to do with it.

The man Toms escorted into immortality at the end of August 1875 found the fame it brought him in equal parts welcome, troublesome and ultimately fatal. Despite being born a long way from the sea in Dawley, Shropshire, when he turned twelve young Matthew arrived in Liverpool with his knapsack and presented himself at HMS *Conway*, ready to become a merchant seaman. Two years later, in 1862, the lad known as 'Chummy' for his amiable nature passed out and joined a Liverpool shipping company running cargo vessels between England, India and China. The fourteen-year-old Shropshire lad found himself sailing around the world.

In the summer of 1872 Webb read of how a twenty-three-year-old Leeds man named J.B. Johnson, widely regarded as the champion swimmer of England, had attempted to swim across the Channel. It wasn't exactly a glorious expedition: he was in the water for little more than an hour before climbing onto a steamer, sailing across to Calais, diving back into the sea and emerging dripping onto the beach announcing loudly that he'd swum from England. The French weren't buying this for one minute, so Johnson quickly changed his story. This was actually a publicity stunt, he insisted, ahead of a genuine attempt that he would be making soon.

Even if Johnson *was* planning an assault on the Channel, his swim never ultimately transpired. The story was enough to plant a seed in the mind of Matthew Webb, however, who after reading of the man's escapade became increasingly fascinated by long-distance swimming. Indeed, he was already proving himself to be quite the athlete. When his ship's propeller became entangled in rope in the Suez Canal, for example, he spent two hours diving down to remove it, experiencing long periods under water with no apparent difficulty. But it was an incident in the Atlantic in the spring of 1873 that truly established him as a swimmer of bravery and heroism.

He was working on the Cunard liner *Russia* when, in stormy seas, a sailor named Hynes fell overboard. As soon as the alarm was raised Webb was plunging over the side to attempt a rescue. The *Russia*, travelling at full speed, was soon far from where he surfaced, but after spotting among the waves what he thought was Hynes he struck out towards what turned out to be just the sailor's hat. Having failed to find the man and watching his ship steaming into the distance, Webb suspected his number was up.

'I had this comfort to carry me to my watery grave,' he wrote later, 'that I had tried to save my fellow man.'

Then he noticed a speck on the crest of a wave a couple of hundred yards away that materialised as a boat launched from the *Russia*, which, having assumed both men to have drowned, was now rowing back towards the ship. Webb roared with all his might, and somehow, above the buffeting of the wind and the crashing of the waves, the crew heard him, turned about and pulled him from a bone-chilling sea in which he'd spent nearly forty minutes. Hynes was never found, but the passengers on the *Russia* raised a purse of sovereigns in recognition of Webb's efforts and when the ship reached Britain he received a gold medal from the Royal Humane Society.

The following year, 1874, Webb announced for the first time his intention to swim across the English Channel, and sought the advice of Robert Watson, editor of the popular magazine *Swimming Notes and Records*. Watson thought Webb incapable of swimming the Channel, partly because he doubted anyone was capable of swimming the Channel, but mostly because he wasn't J.B. Johnson. If Johnson couldn't do it, he thought, this blunt-speaking if amiable young fellow with no verifiable record in long-distance swimming certainly couldn't.

When the two men met in a Fleet Street pub Watson sugared the pill, advising Webb to hold off until the following summer while hinting heavily that he should forget the whole thing and just get on with his life. Webb took Watson's advice at face value, however, and began spending more and more time in the sea at Dover. An early swim saw him cover the thirteen miles between shore and the Varne Bank buoy in four and a half hours, an achievement that attracted the attention of Frederick Beckwith, manager of Lambeth Baths and a coach who at the time was building an impressive reputation. Indeed, Beckwith was impressed enough by a couple of Webb's later swims in the Thames between Westminster and Lambeth Bridges to convince the twenty-seven year old to leave the merchant-shipping service to train full-time under his tutelage.

In July 1875 Webb relocated to Dover, taking lodgings at the Flying Horse, training in the sea for up to four hours a day and drinking at least three pints of beer every night to fatten himself up against the long immersions in cold water. He'd set Tuesday 10 August for his attempt on the Channel, but the day dawned stormy and windy with huge waves crashing over the sea wall around the harbour. The storm lasted all the following day, and it wasn't until a little before 5 p.m. on Thursday the 12th that things had calmed enough for Webb to climb into George Toms' lugger, waiting next to Admiralty Pier, smear himself with porpoise oil and dive in.

Despite a sea that was still lumpier than he would have liked, his head-bobbing, metronomic breaststroke saw him cover nine miles in four hours. Three hours later, however, came his voice from the dark: 'It's no use, there is too much sea on. I must turn it up.'

He might not have made it that time, but in reaching over

halfway to France Webb had swum further across the Channel than anyone on record, easily eclipsing Johnson's half-hearted effort and attracting extra interest in what was surely an inevitable second attempt. He stayed on in Dover, pacing up and down the seafront, staring at the sky and the waves for meteorological hints from the ever-changing conditions in the heavens. He grew nervous and even superstitious, never mentioning the Channel by its name now and referring only to the 'big job' that lay ahead of him. So stressed did he become that on the night of 23 August he apparently committed what a friend described as 'an indiscretion, unusual to him' having 'relaxed his attention to discipline'. A brothel? Quite possibly: Dover was a typical sailors' town and not short of sex workers.

The 24th of August dawned bright and clear and Webb, possibly a little hung over, probably awash with guilt over his mystery indiscretion the previous night, consumed a filling breakfast of eggs, bacon and an entire jug of claret. At a quarter to one that afternoon the crowds gathering on Admiralty Pier to see him off had swelled to such numbers that Toms called him aboard the *Anne* to sail round to the pier's end because it was too crowded to attempt the short walk. He stripped down to his bright-scarlet trunks, slathered himself in porpoise oil, turned to the crowd, pointed theatrically to France, nodded, and at four minutes to one plunged head-first into the water.

Perhaps due to nerves and excitement he set off at a pace much faster than he would have liked, but Toms' advice concerning local tides proved effective and after a couple of hours Webb was already three miles from Dover. As the sun set, fortified by a break for half a pint of beer, he told his support team that he was 'right as a trivet and quite warm'. The light faded into an orange sunset and a clear, starry night. At one point he found himself swimming through a cloud of phosphorescence, his wide strokes sending out billows of luminous green that made a magical sight for those aboard the *Anne*.

By 3 a.m., after thirteen hours in the water, Webb was three miles from the French coast and finally showing signs of tiring. His fatigue was compounded by the turning of the tide, meaning the sea would now be working against him. Things looked so gloomy that at one point the crew prepared to bring him out of the water, but after a brief rest and a cup of coffee laced with brandy Webb rallied a little and struck out again. He'd hoped to land at Cap Gris-Nez, but the tide was pulling him east, back towards Calais. Still he swam on, exhausted, his face pale and waxy, an easterly wind blowing salty spray constantly into his face. So tired now that his legs hung limply in the water, his arms were still thrusting out more by habit than conscious effort.

As the Calais town clock struck ten in the morning Webb found himself barely 300 yards from the beach about half a mile west of the harbour. Word had spread through the town and a large crowd was gathering. Though his eyes were now so swollen by the salt spray they were almost closed, he could still just see figures on the shore in their boaters and bonnets, others running over the sand to join them. As he got closer he could make out individual faces and hear the faint sound of cheering above the gloop and trickle of the sea in his ears. He was so close he allowed his legs to dangle and then he felt it: sand beneath his feet. It took immense effort as his weary limbs were assailed by the full weight of gravity again, but he staggered a few paces out of the surf before collapsing insensible into the arms of his cousin. It had taken a little under twenty-two hours and Toms estimated he'd covered as many as sixty miles – but Matthew Webb had swum the Channel.

He was quickly wrapped in blankets, had hot red wine poured down his throat, then was carried to the Hôtel de Paris where he fell into a deep sleep. Other than waking briefly to consume some soup and some fish, Webb slept for the entire day and the whole of the following night, waking eventually to stiff legs and the pain from the two deep-red chafing welts on the back of his neck, where salt had rubbed between the folds of his skin. He wouldn't be able to fasten his shirt collar for a fortnight.

The hours he spent in that hotel room with its comfortable bed and warm blankets were the last semblance of privacy and normality he would ever know. As he slept, the news of his achievement was already spreading around the world, printing presses across the globe clacking out headlines hailing a super-human feat by an extraordinary man.

As Webb dozed, vaguely aware of the hubbub from the huge crowd gathered outside, most likely the last thing on his mind was what he would do next. Certainly, going back to sea no longer appealed; fame and opportunity now lay outside the door of his room, a chance to earn from his success in a way that skippering a tramp steamer to China couldn't begin to rival. Ultimately, though, the glory of his conquest of the Channel would last for little more than a year. Even by the first anniversary he was reduced to performing cheap stunts and low-key speaking engagements in draughty, half-filled provincial halls. Matthew Webb was many things, but he wasn't a showman. A naturally taciturn, introverted character, swimming was what he did. Wisecracks and cane-twirling were never going to be his thing: as one lecture-goer tactfully noted, Webb was 'not a natural orator'. Even his swimming was hardly the stuff to whip up crowds into hollering frenzies. He was neither particularly fast nor particularly stylish in the water, and certainly no trick swimmer: a steady, rhythmic breaststroke was all he had. Endurance was at the core of his success, and endurance wasn't really the stuff of six nights a week and a Wednesday matinée.

He did his best. When a much-publicised attempt to spend

thirty-six hours in the Thames swimming between Woolwich and Gravesend failed in 1878 due to atrocious weather – and, as Webb put it, he was 'nearly poisoned by the filthy water' – the flimsiness of his appeal as a crowd-puller was exposed. He tried America, taking on challenges such as swimming from Sandy Hook to Manhattan, but the trip proved unsuccessful and even saw him accused of cheating. Similarly, sixty hours swimming around a tank of water at the Westminster Aquarium in London pulled in only handfuls of half-interested people, and while there was a smattering of summer seaside endurance exhibitions Webb's complete lack of showbiz élan worked against him wherever he went.

Five years after his astonishing swim he was thirty-two and in poor health from the constant endurance challenges, and his popular appeal was in terminal decline. He married that year a woman twelve years his junior he'd met at a ball in his honour, but his professional prospects looked bleak and his swimming engagements grew tawdrier.

In 1881, for example, he was forced by the need for income to accept a bet of £100 from a middle-aged doctor named Jennings from Tunbridge Wells for a five-hour challenge in Hollingworth Lake in Lancashire. Whoever swam the furthest in that time kept the hundred smackers as well as the profits from the gate. It wasn't exactly the stuff of showbiz dreams. While Webb was not just the first but still the only person to have swum the Channel, his lack of charisma and the six years that had passed since his landmark swim combined with his unknown opponent and an obscure location to make this even less of an enticing spectator event than competitive corporate accounting. Jennings was, like Webb, an endurance swimmer, a tortoise rather than a hare, and there would be neither tricks nor the thrill of an exhilarating display of racing. Predictably, as the two men entered the water there were just a few knots of people strung along its edge. Webb won, swimming five miles to Jennings' three, but the local newspaper correspondent sent to cover the event conceded that the proceedings 'soon became of the dullest possible description'.

He may have won, but the toll of the cold water was much the greater on the victor. While Jennings hopped out of the water and began chatting to friends, a pale, semi-conscious Webb had to be helped from the lake and into his bathing robe. It took him a week to recover. He was still only thirty-three years old.

Whatever made him decide to swim the perilous rapids at the base of Niagara Falls is anyone's guess, but in 1883, two years after his Hollingworth ordeal, that's what he did. His comment to his wife, 'I don't care, I want money and must have it,' probably provides the best clue, but by this time Webb's mental health seemed to be in decline as much as his physical. He began training close to the Falls through June and July, making announcements of ludicrous plans to swim the notorious rapids every week. Surviving one attempt would be achievement enough; most who knew the rapids regarded Webb's proposed venture as suicidal.

At 4.30 on the afternoon of 24 July 1883, against the advice of locals and the pleas of his wife – and even those of the pilot of the launch he'd hired to take him into the middle of the river – Webb stripped down to his scarlet trunks and dived into the churning rapids, head bobbing, trying to find his rhythm in the boiling surf. A few hundred yards downstream spectators saw him drawn into the outer circuit of a notorious whirlpool, begin to spin with it, raise an arm into the air and disappear below the surface. It was the last time he was seen alive. Four days later his battered body was found in a rock pool miles downstream by a man in a canoe who caught sight of his brightly coloured swimwear among the vegetation.

His inglorious end at just thirty-five should take nothing away from Webb's achievement in becoming the first person to swim the Channel. It was and remains one of the greatest physical feats ever completed by a human being, and the fact it would take twenty-six years and eighty attempts before anyone else succeeded – when the Rotherham-born tyre salesman Bill Burgess crossed in September 1911 – just adds weight to the enormity of that 1875 swim. Nearly all of those whose attempts failed are lost to history now. Even those who made it sloshed out of the waves into a brief flurry of interest, but who remembers them today? Even Bill Burgess is unknown outside a few swimming-history geeks. Although a renowned swimmer, he'd not even trained properly for his swim; he'd happened to be on holiday on the coast at Deal and thought he'd give it a go.

There is one other man in particular among the pioneers who deserves to be remembered, however. He never made it across the Channel but, by Neptune, it wasn't for want of trying. Jabez Wolffe was a persistent and consistent failure as a Channel swimmer, but that didn't stop him becoming the most famous of his day; he was also a successful and opinionated coach and a fount of Channel knowledge. What George Toms was in a wheelhouse, Jabez Wolffe was in the sea itself.

Wolffe had all the attributes of a Channel swimmer. A barrel of a man, photographs show him to have had a chest of such dimensions that he could have taken out a copse of trees with a single cough. His head was so big his hair couldn't keep up and looked like a toupee on a beach ball. So large was Jabez Wolffe that when he got into the Channel it's a wonder that Archimedes' principle didn't trigger the submersion of Dover and Calais.

For all his contemporary fame, little is known of Wolffe's origins beyond a few brief notes gleaned from his *Textbook of Swimming*, published at the height of his popularity in the 1920s (it's a slim volume but comprehensive in its instruction – its opening chapter advises the would-be swimmer to begin by sticking their head into a wash-basin of cold water). He was born in Glasgow in 1876 to a Polish-Jewish jeweller, and he demonstrated from an early age the stamina and strength on which he would make his reputation. He relates in the book a story of winning, as a youngster, a long-distance race in a public baths, then taking a central role in a water polo match straight afterwards and leaving the pool feeling he still had plenty in him. On his way to the changing rooms he was collared by 'Mr George White, the champion one-armed swimmer, who was among the company present' and who told him that on the evidence of what he'd just seen Wolffe stood as good a chance as anyone of emulating Matthew Webb by swimming the Channel.

That conversation would change Wolffe's life, turning it into one dominated by the English Channel. In July 1905 he did his first major long-distance swim in the Thames, from Blackwall to Gravesend pier in a little over five hours. Three weeks later he swam from Dover to Ramsgate in four and a half hours and a month after that covered the thirty-four miles between Margate and Herne Bay in nine and a half. Both those latter swims he made with Annette Kellermann, a well-known Australian actress and renowned long-distance swimmer, who pioneered the one-piece bathing suit for women despite being arrested for indecency while wearing one in Illinois.

By the summer of 1906 Wolffe was thinking seriously about having a crack at the Channel. He made another swim from Dover to Ramsgate, this time entering Ramsgate harbour itself as Matthew Webb had done while training for his successful Channel swim. Wolffe reached Ramsgate in no time, but then spent fully two hours battling the tides and eddies at the harbour mouth, his resilience aided by his friend Major Nicholl who kept his spirits up and his rhythm regular by playing the bagpipes. Wolffe had noticed that pipe bands helped to keep marching soldiers moving briskly and in step and considered there to be no reason why the same should not apply to swimming.

'He kept skirling away,' Wolffe wrote of Nicholls' contribution, 'and announced to those with him in the accompanying lugger his fixed determination to blow me inside the harbour.'

Whether blown in by a piping major or not, Wolffe's persistence paid off and he made it after an almighty struggle, one that left him convinced he could not only swim the Channel but do it inside two tides.

Jabez Wolffe would never succeed in swimming the Channel but nobody tried harder. Most people would give up after one failure; Wolffe tried twenty-two times. 'I certainly ought to have succeeded the first time,' he rued, thinking back to 18 July 1906 when his fateful relationship with the Channel officially commenced. He'd set out in excellent weather conditions in a calm sea and was going well, clocking up fifteen miles in seven hours, but this caused him to lose the run of himself and, well, get a bit cocky. 'I felt very pleased and elated, full of strength and vigour,' he recalled, 'and I suppose that I succumbed to my feelings. Anyway, I must have spurted up tremendously for just after this my left leg gave out.'

Wolffe was less than four miles from France and had been swimming for less than ten hours before he, erm, spurted up tremendously, and at that rate he was set not only to succeed but to finish in practically half the time it had taken Webb thirty-one years earlier.

'I can't help thinking even now that Fate served me a scurvy trick by making me pay so heavy a penalty for one or two exuberant, but unwise, kicks of joy,' he grumbled. Scuppered by his joyful kicking and tremendous spurting, he would try again thrice more that summer, twice before his leg had had the chance to heal and once when he was 'overcome by biliousness', at which point his attempt ended.

In September 1908 Wolffe got as close to the French shore as he ever would: there were just 800 yards between him and success before the tide turned and carried him gently but firmly back out to sea. If it hadn't been for a thick fog, he insisted, in which he'd lost his bearings mid Channel, he would have made it ashore on the tide.

'A fifteen and a quarter hour swim against a strong sea and wind and a sprained wrist and a close view of Calais were all I had to show for it,' he sighed.

A twenty-two-mile jaunt from Eddystone Lighthouse to Plymouth in 1914 gave him a bit of a change: a swim accomplished in less than eleven hours that may have been prompted by the presentation of an illuminated address by the Corporation of Devonport in recognition of his 'many plucky attempts to swim the Channel'. That phrasing must have stung Wolffe like a jellyfish – something that also prematurely ended more than one of his Channel attempts, as it happens – because there's no more patronising compliment than 'plucky'. Fortunately, he was never one to let something as inconvenient as persistent failure damage his colossal self-esteem. In the autumn of 1906, barely two months after his first failed attempt, he was already giving demonstrations as 'the famous Channel swimmer', when at that stage he was in real terms no more a Channel swimmer than I am. Nevertheless, he always ensured his attempts to cross to France were carried in newspapers across Britain, Europe and even America. In 1907 he starred in *The Channel Swimmer*, a short film showing him being greased up and entering the water seen in picture houses across the country. Despite every attempt ending with a grumpy Wolffe hauling himself dripping and complaining into a boat somewhere between Dover and Cap Gris-Nez, he was successfully turning himself into the biggest Channel celebrity since Matthew Webb. Webb's mistake was in succeeding, leading to a life of unfulfilled misery. Wolffe, however, succeeded by failing. Repeatedly.

In the great British tradition of heroic failures he was right up there with the best of them, a greased-up England penalty taker, an Eddie the Eagle in trunks. Wolffe became a familiar sight to Dovorians each summer, puffing out his forty-seven-inch chest and announcing that this time, *this* time, he was sure he was going to do it. There would be a great hullabaloo at the quayside, and as he splashed off towards the horizon to the fading skirls of Major Nicholls' bagpipes, people would turn to each other and speculate over what would thwart him this time. There was that occasion he was done for by a piece of floating timber, they'd recall, and the shark that apparently butted him amidships and put him off his stroke (something Wolffe regarded as 'a cowardly act', and 'bad sportsmanship' on the part of the shark). Then someone would shout 'Ah, here he comes' as a vessel appeared in the harbour with Wolffe, his huge head red and puffy with rage and exertion like the world's fattest tomato, stomping about on the deck with a blanket round his shoulders railing at the injustice of it all and already planning his next go.

He might never have swum it successfully, but no swimmer in history has spent as long in the Channel as Jabez Wolffe. What he didn't know about tides and drifts and eddies and jellyfish and what to eat and how often wasn't worth knowing. The only thing he couldn't advise on was what to do when you actually got to the other side.

He made his last attempt in 1921 at the age of forty-seven, shortly after writing a piece for a local Dover newspaper entitled 'Is the Channel Swim Worthwhile?', a question that he was more qualified to answer than anyone. Later, on the eve of the Second World War, he held forth to an interviewer about his Channel experiences.

'I swam 600 miles in the Channel during twenty-two attempts,' he mused. 'On one occasion I swam forty miles trying to cross the Channel and missed the tide at the end of it by five minutes. I'd have done it that day in July 1914, too. Sea like a mill-pond, glass high, tides at their best, the kind of day I'd been waiting for for years. Then came the order from the French authorities – war had broken out and we had to race home by steamer at once.'

Poor old Jabez. Finally all the stars align for a successful Channel swim and blow me down if a global war doesn't break out while he's in the water. It's almost like they did it on purpose.

'Adventures?' he continued. 'Lots of them. Was nearly run down by a liner whilst swimming at night. Swam through a shoal of mackerel and thought I'd go mad the way they tickled me. And then of course there was the time I felt shuddering below the water and my teeth started chattering. The sea was vibrating. Couldn't make it out until they shouted a warning from the boat. I changed course and spurted just in time to avoid a submarine surfacing.'

After failing in his final attempt, Wolffe turned his attention to coaching young aspirants, notably the women swimmers who came into their own during the 1920s. Its pioneers may have been male, but one thing that sets Channel swimming apart is that as many women attempt the swim as men and just as many succeed. Channel swimming is a feat of human endurance in which gender is entirely irrelevant.

It was in the aftermath of the First World War that a new generation of independent women began to set their sights on the Straits of Dover. With men away at the Front, women had effectively kept the country running during the war, working in factories and workshops, keeping up the production of munitions and wider heavy industry. The success of the suffragette campaigns demonstrated the new determination of women to push back fiercely against their male-dominated world. What greater demonstration of equality could there be than swimming the Channel?

The first attempt by a woman had been as long ago as 1900, when Baroness Walpurga von Isacescu of Vienna showed up in Calais with a bathing costume and an ambition. As decorum dictated, she entered the Channel from a bathing machine outside the Calais Casino at 7.30 a.m. on 21 September, after breakfasting on two eggs and a glass of port. A couple of miles out into the Channel the wind blew up and impeded her progress, carrying her to the east, but when the tide turned in the early afternoon she was carried back towards Dover. By 5.30 p.m., however, the temperature had dropped and the Baroness had started to struggle. It took some persuasion but eventually she agreed, on her pilot's advice, to abandon the attempt. She'd been in the water for ten hours and swum more than twenty miles.

Other than an attempt by Kellermann in 1905, the Channel was a male preserve until after the First World War. Once the Channel had been cleared of mines there began first a trickle and then a flood of women attempting the crossing. It wouldn't be long before one of them succeeded, either.

Manhattan-born Gertrude 'Trudie' Ederle was already a swimming superstar when she arrived in France in 1925 for her first attempt. Barely eighteen and the daughter of a German butcher, her incredible speed through the water had already earned her a number of accolades. Between 1921 and 1925, the year she turned professional, Ederle smashed twenty-nine world swimming records, breaking seven in one remarkable afternoon. Although she was hot favourite to win Olympic gold in the Paris games of 1924, the combination of a knee injury and the US team having based itself a five-hour round trip from the capital so as to keep the athletes safe from the City of Light's 'bohemian morality' restricted her to two individual bronzes and one relay-team gold.

When she first arrived in Dover she had just set a record for swimming between Manhattan's Battery Park and Sandy Hook, accomplished in a time of seven hours and eleven minutes that would remain unbeaten for eighty-one years. But this was merely a warm-up for an attempt to become the first woman to swim the English Channel. To help achieve this the US Women's Swimming Association engaged as her trainer none other than Jabez Wolffe.

Wolffe was bullish about Ederle's chances before she'd even arrived in the country, predicting that she would not only succeed but shave a hefty chunk from the fastest recorded time, fourteen hours and twenty-five minutes, by the Italian swimmer Enrico Tiraboschi two years earlier. Only five people had swum the Channel successfully when Ederle arrived in France, so Wolffe's confidence either displayed carefully reasoned optimism or was expressed while he was still negotiating his fee. By early August, with Ederle's attempt just a couple of weeks away, he had changed his tune.

'I think Miss Ederle is the fastest swimmer alive but I do not expect her to succeed in swimming the Channel this time,' he said of the teenager. 'She refuses to train and plays the ukulele all day. Apparently she broke her records without much training and doesn't see why she should train now. Also, she swims too fast.'

As it happened, Wolffe's prediction was right on the money. Ederle began at Cap Gris-Nez early on the morning of 18 August and made good progress until mid afternoon when a westerly wind blew up. She began to struggle as the waves grew larger and the giant Egyptian swimmer Ishak Helmy, who'd been assisting Wolffe while preparing for a Channel attempt of his own, dived into the water alongside her to check on her condition. When he reported back to Wolffe that the American was close to collapse with six miles still to go before Dover, Wolffe told him to pull her out of the water. She was barely conscious: any longer, surmised the Channel veteran, and the American superstar would have drowned.

By the time she'd returned to the USA in mid September Ederle had decided the trainer was to blame for her failure. Wolffe, it was claimed, had ordered her out of the water prematurely and against her will. Wollfe, unsurprisingly, wasn't taking that lying down.

'Her statements are quite untrue,' he thundered from his home in Brighton. 'I take it her story was meant to cover her non-compliance with my prepared efforts to get her to train. It was evident to both English and French observers that her training consisted mainly of sitting about playing her ukulele.' Musical sniping was a bit rich coming from someone who had a bagpiper follow him back and forth across the Channel, but Wolffe was only getting started.

'I am of the opinion that had Miss Ederle followed my instructions and not been interfered with by the American advisers accompanying her she would have succeeded in swimming the Channel. I still think she is capable as long as she is properly trained,' harrumphed the man employed to properly train her.

'Of course I played my ukulele,' responded Ederle from across the Atlantic. 'But only in the evenings when my training was concluded. It was my only relaxation.'

The following year she was back, a little older, probably wiser, and with a new trainer – this time someone who had succeeded in swimming the Channel, one Bill Burgess. On 6 August 1926 Ederle was up before dawn at the Hôtel la Sirène near Cap Gris-Nez to demolish a bowl of cornflakes, half a fried chicken and a plate of bacon and eggs. As she threw down her knife and fork onto an empty plate, Burgess asked how she was feeling.

'Like I could lick Jack Dempsey,' she replied.

A little after 7 a.m. Trudie Ederle waded into the sea and set out for Dover, accompanied by her support boat the *Alsace* with Burgess and her parents among those on board. She covered two miles in her first hour despite a brief stop for stomach cramps and by late morning she was going well, six and a half miles north of Cap Gris-Nez and singing along to the gramophone records being cranked out by her mother on the *Alsace*.

'The water is wonderful, I could stay in for a week,' she chirruped.

At midday she rested briefly, drank a baby's bottle filled with beef tea and ate some chicken. In the middle of the afternoon her pace dropped, the wind rose, it began to rain and her chirpy commentary stopped. She kept swimming but the cheeriness had gone, along with the good weather. Ederle's team attempted to keep up her spirits with a lusty rendition of 'Yes, We Have No Bananas', which must have helped, but then the tug hired by rival newspapers to the New York *Daily News*, with whom Ederle had signed an exclusive agreement, turned up and proved to be a more serious danger to her success than wind and fatigue. As photographers strained for close-ups, twice the vessel veered too close to the swimmer, throwing bow waves that submerged her altogether.

Another brisk westerly wind blew up. The *Alsace* skipper told Burgess and Ederle's father they were being carried too close to the notorious Goodwin Sands and would need to take a new south-westerly course to avoid them. This would add significantly to the distance Ederle needed to swim but their current course was too dangerous, he said. The sea rose further, substantial waves breaking over the bow, while for worryingly long periods Ederle's bright-red swimming cap kept vanishing between the steepling waves. Skipper, father and trainer pored over the charts in the wheelhouse and realised that once she made it past the South Goodwin lightship the sea would become calmer and the tide would be in her favour. Burgess went out and gave her the news, looking down from the heaving ship at the red speck rising and falling with waves much larger and more powerful than any he'd experienced during his own Channel swim. Seeing this vulnerable young woman pressing on in such dreadful conditions, Burgess was almost overcome with admiration.

'God almighty,' he called to nobody in particular as he clung to the rail, the wind whipping the words from his mouth, 'I never saw anyone so marvellous!'

The Channel wasn't done with Trudie Ederle yet. So severe had the wind been that when the skipper checked his chronometer he found that the tide was turning almost two hours early. They were now a mile from the English coast, but that mile would be the toughest of them all. Ederle would need to turn away from Dover, they realised, and aim instead for St Margaret's Bay near the South Foreland Lighthouse. Burgess told her the news as he handed over a couple of slices of pineapple, and Ederle bore it stoically, taking a deep breath and throwing herself towards the shore. Her warhorse of a trainer was still utterly in awe.

'No man or woman ever bore such as swim,' he said as he turned back to the rest of the crew. 'It is past human understanding.'

Half a mile from the shore, at about 7.45 p.m., Ederle began for the first time to doubt if she could make it. Then she looked up and saw a crowd of people gathering on the shore outside the Zetland Arms inn at Kingsdown, just south of Deal. They were lighting flares on the beach to help guide her home. Burgess and her father climbed into a rowing boat and accompanied her over the final few hundred yards, Burgess counting them down as they went, Ederle's red cap still disappearing from view beneath the foam for seconds at a time, until eventually her feet touched shingle. After fourteen and a half hours in the water, a full two hours faster than anyone ever before, she felt suddenly heavy as she stumbled through the surf and was engulfed by the cheering crowd. Her father retrieved the bewildered, exhausted swimmer from the well-wishers, returned her to the *Alsace* and made the short journey to Dover, where a comfortable bed awaited her in the Grand Hotel.

An estimated two million people lined the streets of Manhattan on her return, where she was granted a ticker-tape parade along Fifth Avenue. The city's official greeter Grover A. Whelan said that of all the greats he'd welcomed and escorted on their parades none had made the impact Trudie Ederle did. President Calvin Coolidge described her as 'America's best girl'. With her bobbed hair and sparky personality, she had proved that she could not only match men but outperform them at one of the world's greatest tests of physical endurance. She became a symbol of the modern, liberated American woman. Like Webb before her, however, she found fame hard to handle, and having been bundled onto a vaudeville tour and then into a Hollywood film of her story, not to mention the hundreds of marriage proposals she received, everything finally got too much.

'I got the shakes,' she told an interviewer years later. 'I was just a bundle of nerves. I quit the tour and by the time I did I was almost completely deaf.'

Her hearing had been damaged by a bout of measles as a toddler, a condition exacerbated by her swimming and worsened considerably by the battering she'd taken in the Channel. The added stress of relentless adulation had only made the condition worse until she could hear practically nothing. She became engaged in 1929 but when she told her suitor he might find it difficult having a wife who couldn't hear, he conceded she had a point and promptly vanished.

'There never was anyone else after that,' she said. 'Anyway I wanted to avoid the hurt.'

In 1933 Ederle slipped on some broken tiles in the stairwell of the building where she lived, and injured her back so badly it was feared she would never walk again, let alone swim. Bedbound for four years, she managed to recover enough to make a brief appearance at a 1939 aquatic circus held as part of the New York World's Fair, and during the war worked testing aircraft instruments at LaGuardia Airport. She spent many years teaching deaf children to swim and lived to the age of ninety-eight, making it into the new millennium – she died in 2003.

'I have no complaints,' she said in an interview towards the end of her life. 'I am comfortable and satisfied. I am not a person who reaches for the moon as long as I have the stars.'

Back in the Channel, it wasn't long after Ederle's epic swim that Britain had its own Channel-conquering woman. Mercedes Gleitze was born in Brighton in 1900 to German parents Heinrich and Anna, but with her mother struggling to cope with three daughters she spent much of her childhood with grandparents in Bavaria. She returned to Sussex in 1910, but four years later when war broke out Heinrich was arrested and interned as an enemy alien, prompting Anna to take herself and her three children back to Bavaria.

By 1918 the war was over and Mercedes was eighteen, headstrong and yearning to return to England. When her mother refused to take her back to Brighton she promptly set off on foot, penniless, walking more than 400 miles north to the German coast, where she waded into the North Sea and began swimming in the direction of England. Although she was a strong swimmer, with no knowledge of the tides or experience of sea swimming beyond a few childhood dips in Brighton, she began to struggle. After hours in the water she washed up semi-conscious on the island of Wangerooge, the easternmost inhabited island of the Frisians, where she was found and taken in by a local man named Hermann Rosing and his daughter. She was, they told her, incredibly lucky to be alive: if she'd missed the island the tide would have carried her out into the North Sea and certain death.

Even this news wasn't enough to dampen Gleitze's determination, and after resting for a couple of days she thanked her hosts and prepared to return to the sea. Luckily, before she could embark on her suicidal plan her mother arrived. Intense negotiations followed in which Mercedes secured guarantees of greater independence and the right to earn her own living until she could save enough to move to England. Grudgingly she accompanied her mother back to Bavaria.

By the time she turned twenty she had saved enough to move, taking the more conventional train and boat this time. She found work in London as a stenographer, joined a swimming club in Holborn and spent summer weekends swimming in the Thames and in the sea at Folkestone. Whether it was being able to see the coast of France from the water or a feeling of unfinished business with the sea after her Frisian experience, in August 1922, four years before Ederle's success, Gleitze made her first attempt to swim the Channel, a shoulder injury curtailing her attempt after six hours.

She still fared better than the French-Canadian swimmer Omer Perrault, the unluckiest Channel swimmer of them all, who made his own attempt on the same day and also had to abandon his swim after he got into difficulties just three hours in. Returning the following year, he was pulled into his support vessel after twelve valiant hours and taken below to warm up by the stove. A member of the crew decided it would be a good idea to sponge the insulating grease off Perrault's body with petrol as he rested beside the hissing, spitting stove. Unsurprisingly, the swimmer went up like a brandy-soaked Christmas pudding. He ran screaming up to the deck intending to chuck himself back into the Channel, but his quick-thinking wife grabbed him as he passed and managed to smother the fire with her overcoat. The thickness of the grease prevented serious burns, and after one more attempt in 1926 when his support boat turned up in the wrong place and then went home, he accepted that the Channel Fates would never favour him.

By 1927 Trudie Ederle had become the first woman to cross the Channel, and the same kind of stubborn streak that had compelled Gleitze to take to the waters drove her on. When she stood on the shore beneath Cap Gris-Nez at three o'clock in the morning on 7 October 1927, much later in the year than any other swimmer, she had seven failed attempts behind her and her eighth was looking the least promising of them all. It was a foggy, chilly night, the sea was cold and the hours ahead looked pretty grim, especially as she could still feel in her limbs her previous attempt, a five-hour effort scuppered by heavy seas made just two days earlier.

Despite fog so thick that for most of the crossing neither she nor her support boat were entirely sure where the other was – not to mention the fact that nobody had ever made a successful crossing later than mid September – an exhausted Gleitze came ashore near St Margaret's Bay fifteen hours after setting out. She'd survived the ordeal fuelled by nothing but tea, grapes and honey, prompting the Press Association's correspondent to declare her 'the most amazing girl in England'.

That should have been that, a scatter of interviews, a civic reception, a couple of lucrative endorsements, maybe a trundle round a few theatres and music halls, but when a week after the swim Dr Dorothy Logan appeared out of the waves at Folkestone claiming to have swum the Channel in a record time of just over thirteen hours, Gleitze's problems began. Now, Dr Dorothy Logan was at it – a phoney, a fraud, a hornswoggler who had no more swum the Channel than you or I. She wasn't even very good at fronting up her flimflammery: when a few vaguely probing questions were posed about her swim she folded instantly, admitting she'd made the whole thing up. But, she said, wholly unconvincingly, that was actually the plan all along: the whole caper was designed to show how open to abuse Channel swimming was.

You'd think Logan being rumbled almost before she'd finished towelling herself off would prove that Channel swimming wasn't really open to abuse after all, but instead the episode shifted the focus onto Gleitze's swim the previous week. Here, after all, was another woman who'd claimed a fast and successful swim long after all the other Channel swimmers had packed up and gone home. Even the Channel Swimming Association raised a quizzical eyebrow in her direction and requested more detailed documentation. Stung by the criticism and grievously offended by the suggestion that her swim had not been above board, Gleitze characteristically went one better: stuff the documents, she'd just swim the Channel again. What became known as the 'vindication swim' not only gripped the British nation but reached across the Atlantic, where the Rolex MD

Hans Wilsdorf sent Gleitze one of their new waterproof Oyster watches to wear on her swim.

By 21 October the temperature in the Channel had dropped to 13°C, ordinarily far too cold to attempt a crossing, but at 4.20 that morning Gleitze was back at Cap Gris-Nez, slathered in grease and entering the water for the third time in three weeks. The conditions she faced were all but impossible, with strong winds and steepling seas from the start. After eleven hours she was still eight miles short of the English coast and barely conscious when her support crew, concerned that their inquiries for the last hour had been answered with 'Just let me sleep' and aware they'd just missed the tide needed to carry her to Dover, pulled Gleitze out of the water in a state of near-delirium. The guitarist and banjo player on board who had been playing constantly to keep her spirits up also stopped, adding to the general feeling of relief.

The world went nuts for Mercedes Gleitze. Yes, she'd failed in her 'vindication swim', but she'd given it a right old wallop. It was another perfect British story: a glorious failure that felt, in its self-flagellating, righteously indignant way, like a glorious victory. The saga brought fame, attention and a flurry of marriage proposals. Mercedes handled those like she handled most things: idiosyncratically.

At the time of her Channel heroics she'd been engaged for nearly ten months to a man she'd never met. Most female Channel swimmers reported a hefty volume of correspondence and attention from men with matrimony in mind – the mixture of fame, tenacity and calves that could crack walnuts setting a significant percentage of the male population all of a tizz. Mercedes Gleitze, with her blue eyes, long curly hair and dimpled smile, set so many hearts aflutter with her Channel exploits that a tangible draught swept the country. But alas! To the disappointment of the large section of the British male population smoothing down their hair and nervously fingering their hatbands in anticipation, Mercedes Gleitze was already spoken for.

In December 1926 she'd received a letter from a soldier named William Farrance stationed with the East Lancashire Regiment in India who'd seen her picture in a magazine article and thought, well hello there. 'It was an awfully nice letter,' she said, 'and so different from the thousands of other letters I receive from men.'

In the article Gleitze's advocacy of social welfare was mentioned – she'd use some of the money that came with her Channel fame to open a homeless shelter in Leicester – something that also interested Farrance. He was, he said, quite happy to help out with the work she wanted to do and if they had that in common, well, wouldn't it make sense for them to get married? A pretty swift escalation, but its recipient wasn't fazed.

'At first I was inclined to ignore the letter,' said Gleitze, 'but after re-reading it and thinking it over, well, I simply couldn't get it out of my mind and in the end I yielded to his request.'

At this point she hadn't even seen a picture of her fiancé, who only sent her a photograph of himself once she'd said yes. Even then it would be two years before the couple finally met, in November 1928 on the steps of Westminster Abbey. The half-hour stroll around St James's Park that followed, however, was enough to make them aware that marriage probably wasn't such a doozy of an idea after all. Gleitze let Farrance down gently, blaming her swimming ambitions for getting in the way of her heart.

'He is an exceedingly nice man,' she said, 'but I do not consider myself fit to be any man's wife because of my passionate love for the sea. I am convinced that I shall never be able to settle down as a wife until I have swum the Irish Channel, the Wash and the Hellespont.'

Eventually she married an engineer from Dublin called Patrick Carey, another starstruck fan who had followed her all over the country watching her exhibitions. The wedding took place in a Dover church in August 1930, after which the newly-weds immediately set out for Turkey where Gleitze was to swim the Hellespont. Bill Farrance may have noted the order of those events with a wry smile.

Her Channel swim had triggered a busy and lucrative career undertaking high-profile water-based challenges, but when in the spring of 1928 she conquered the Strait of Gibraltar certain quarters tried to cast doubt on that achievement too. Despite two fishing boats accompanying her from Tarifa in southern Spain to the Moroccan coast, a twenty-four-mile swim that she accomplished in just over twelve hours, *The Times* noted that 'there appear to have been no English witnesses other than a small boy', and that while plenty of Spanish people confirmed the swim they'd all talked at once, very fast – and most underhand of all – in Spanish.

In addition to the Hellespont, Gleitze swam 100 miles around the Isle of Man and travelled to Australia and New Zealand for marathon swims, crossing the strait between the North and South Islands in the latter, and in South Africa she became the first person to swim from Cape Town to Robben Island and back. Then having worked tirelessly for seven years after her Channel crossing, in 1934 she suddenly withdrew from the public eye shortly before the birth of her first child. Indeed, so quietly and anonymously did she live with her family in Wembley, it was only after her death in 1981 that her daughters learned of their mother's erstwhile celebrity, turning up some old scrapbooks when clearing the attic.

As for Jabez Wolffe, he died in 1943 and his death made the news all over the world. Gertrude Ederle's record still stood, which must have irked him given the fractious nature of their relationship. Never short of an opinion, Wolffe had in a 1939 interview confirmed that he would always be happy to train someone capable of beating the record, as long as she was the right kind of girl. London women wouldn't do, for a kick-off.

'I would have nothing to do with those because training for the Channel is too monotonous for London girls,' he sniffed. 'The training lasts for eight months and includes 1,500 miles of walking, 200 hours of swimming and 100 hours of massage, not to mention many other hours of exercise. Before the training is halfway through the London girl is crying to get back to what she calls civilisation. People who are going to swim the Channel have to go to bed early and my wife will tell you there is nothing more difficult than getting a London girl to bed before 9.30. That is just the time when most of them are beginning to wake up for a few hours of what they apparently regard as enjoyment.'

As for the kind of woman he would be willing to train, it seemed hair colour was the key.

'She would have to be blonde,' he insisted. 'Genuinely blonde. Brunettes are not stickers, they get nervous, they give in. I remember one blonde-haired girl who started pulling exhausted faces on a Channel swim and the French people on the boat raged at me for not bringing her out of the water. They called me a brute; two women even spat at me, and when at last I called to the girl to come in the moment she heard my voice she sprinted through the water as fresh as you like. Blondes are calmer, more stable, better stayers.'

I thought about this in the sea at dawn this morning. France was clearly visible across the calm, dark-blue water as the impending sun sugared the thin clouds with a pink frosting. It looked closer than ever. Now that I've learned a little from the pioneers about what it takes to be a Channel swimmer I can combine it with what I know about tides and the weather and, most crucially of all, I have blond hair.

See you on the other side.

10

Some Channel People

One morning at the end of January I got up, put on my swimming shorts, little rubber sea shoes and fleecy robe, made my mug of tea, and opened the door to find it had snowed during the night. I wasn't exactly emerging into a blizzard but there was enough to cover the ground, and when I reached the beach there was a light layer that undulated with the shingle to within a few feet of the waterline. According to the weather app on my phone the temperature had dropped to minus 3°C, which it rounded down to minus 5 to include the wind chill. It was a beautiful morning and there was a huge mass of fluffy clouds to the south, all pink-hued and orange at the edges in the light from a sun that was still just behind the horizon. I could have quite happily stood there wrapped up warm, sipped my tea, watched the sun come up and gone inside again.

The worst part of swimming in the Channel on cold mornings is not getting into the water but getting out of your clothing. I was toasty. It was a glorious, if bone-shudderingly cold morning. Why would I want to strip down to a pair of shorts and wade into the sea when the temperature was well below freezing? Ordinarily I wouldn't be out in this cold at this time even dressed like an Arctic explorer, let alone practically naked. I braced myself, grabbed the zip, pulled it down and slipped out of my robe, letting it drop onto the snow. The cold hit me so hard I almost heard it, a shock great enough to dislodge the memory of a poem I learned at school when I was seven and hadn't thought of since. I was accustomed to cold mornings now but this, this was the coldest yet.

I bent down, reached into the pocket of my robe, pulled out a pair of neoprene gloves and stretched them onto my hands. A few days earlier on a similarly cold morning I'd done something particularly stupid: finishing my swim, coming out of the water, picking up my phone, then going back into the water and recording a birthday video message for a friend of mine who had refused to believe I was still swimming in the Channel in January. I'd sloshed around for a bit, mugging and messing, hit the off button and waded out of the 7-degree water again, all pleased with myself.

I had, however, been in the water far too long. The pain started before I'd left the beach. My feet were numb and my fingers progressed quickly from a sharp tingle to a deep burning pain as if I'd stuck them in a mincer. I got through the door and ran my hands under the cold tap, knowing that warm water would only make things worse. After a few minutes the pain had been replaced by a persistent throb, then numbness, but by the time I got into the shower the feeling was returning to all my fingers. Except one. The tip of the ring finger of my left hand stayed numb. It would go on being numb for about three months before the feeling slowly came back, by which point I'd convinced myself I'd done permanent nerve damage. It was exceptionally stupid of me. Hence the gloves, the kind scuba divers wear. I'd resisted buying a wetsuit not out of any kind of macho bravado but the opposite: I knew if I started wearing one I'd never stop, and in the height of summer people would be frolicking in the waves wearing speedos and bikinis while I was still tentatively toeing the water dressed neck to ankle in black neoprene, like some kind of lumpy, cowardly Milk Tray man.

Wading into the water when it's really cold is actually more straightforward than you might think. The sea at that time of year tends to be warmer than the air, and while it doesn't necessarily feel like it, there's not the same level of shock as there is when you enter the Channel in the summer. That's not to say it was like a dip in the Caribbean, of course. Hoo, boy, no. As the water level rose up from my midriff I realised I was shouting so loudly there were probably people in Dunkirk looking up from whatever they were doing, but finally my shoulders were in and I was swimming, if you can call thrashing your arms around frantically while gasping *ohmygodfuckmefuckinelljesusgodshittingnora* swimming.

It wasn't long before I calmed down a bit and began swimming more like a human and less like an octopus playing the drums, when I looked back at the beach and it struck me that I was in the English Channel on what was sure to be one of the coldest days of the year, swimming, in shorts, as the town was waking up to snow. I turned away to swim out a little further and saw the top of the sun peep over the horizon. Treading water, I watched as it became an orange semicircle that launched a golden, shifting carpet across the surface and a fiery orange path between it and me that stretched all the way to my chin. The Channel had transformed from a dullish grey-green to this shimmering, delicate, shifting meniscus of deep, ever-changing colour and light, and I was surrounded by it.

In that moment, the time it took for the crown of the sun to appear to when it detached itself from the horizon, I was suddenly aware of the Channel like never before, aware of all the people that had travelled along it and across it and those who'd never left it. This was a stretch of water connected to more people, probably, than any other in the world, and here I was, one of them, a Channel person. I looked around at the snowy shore. I could see the chimneys of the houses on the front but that was all. It was just me, the Channel, the shingle bank, the horizon and the sky. I was alone with the Channel and alone with all the stories of all the Channel people.

Early in the morning of Sunday, 29 August 1920, nineteen-year-old Maurice Braddell and his friend, the euphoniously named Adelin Eugene Paul Firmin Marie Ghislaine van Outryve d'Ydewalle, carried the two-man canoe they'd hired down the beach next to Folkestone's Victoria Pier, placed it in the water, climbed in and set out to make the crossing to France.

Both were, by any definition, toffs. Braddell was the son of Sir Thomas de Multon Lee Braddell, a former Attorney-General of Singapore and Chief Judicial Officer of the Federated States of Malay who had retired to Folkestone in 1917. D'Ydewalle, known as 'Diddles', was the son of a Belgian count who had like many thousands of Belgians fled to Britain to escape the war. Adelin's sister Suzanne would become the grandmother of Mathilde, the present Queen of Belgium. So yes, these were not, by any stretch, horny-handed tillers of the soil.

Both men were of the first generation to have been too young to serve in the Great War and perhaps it was that mixture of regret and good fortune that fired their sense of adventure. They'd told friends and family they were planning to row to Dymchurch, a little further along the coast, but when they were seen making for the horizon it was clear they had a more ambitious destination in mind.

While conditions near the coast had been calm, out in the Channel the sea was cutting up rough enough to provide the sternest challenge even for experienced oarsmen like Braddell, who had rowed for his Oxford college, and when there was no word of or from the men by Monday evening the situation looked grave. The families awaited the arrival of every steamer from Boulogne and watched hopefully as the crowds of people disembarked, hoping in vain to see their offspring among the forest of hats.

As Tuesday dawned, with no news or sightings in forty-eight hours, the families were giving up hope. At lunchtime, however, a French fruit steamer docked at Folkestone from which disembarked two dishevelled, slightly sheepish teenagers with a story to tell.

'We rowed steadily for eight miles until one of Maurice's oars broke,' d'Ydewalle revealed.

They'd coped with three oars for a while, angling carefully against the tide and were still confident of making the crossing, but then the tide turned and sea conditions grew rougher in a rising wind. Then fog descended, through which they just about made out the glow of the Varne lightvessel about three miles distant, and decided to make for it even as the tide ran harder and harder against them.

'When we got to the lightship the crew took us on board and made us very welcome and comfortable,' said d'Ydewalle. 'They told us they wondered how we got so far in the conditions and said they would not even try it in one of their lifeboats.'

The pair decided to abandon their crossing and stayed on the ship waiting for a vessel that might take them home. The crew of the lightship signalled to the passing traffic without much success – one Gravesend-based tug came alongside but declared itself unable to accommodate passengers – until the French boat agreed to take them home.

'We spent the time on the lightvessel fishing and playing games,' said the Belgian. The pair talked big about having another go at the Channel but this was probably youthful bravado. For a start, their parents wouldn't be letting them anywhere near a set of oars for a while.

After they'd landed at Folkestone the friends' lives took different paths. D'Ydewalle, who had taken British citizenship, was a cadet at the Sandhurst military academy, from which he was commissioned into the Army in December 1920, joining the South Staffordshire Regiment as a second-lieutenant, then posted to Ireland. In his first summer as an officer he was involved in the deaths of two suspected IRA men in Cork, in separate incidents. D'Ydewalle, who had a reputation for brutality against prisoners he suspected of being IRA men, insisted both had been trying to escape after being arrested, an account disputed by locals to this day. He married a local woman in Cork, left the Army in 1930 and set up a car-hire business in Hampshire which went bankrupt in 1936. He died in a Penarth hospital in 1943 at the age of forty-two.

Braddell, meanwhile, was already an actor of some promise when he went missing in the Channel. Despite coming from a long line of lawyers, after the aborted crossing he enrolled at RADA and enjoyed some success on the London stage during the 1920s. Towards the end of the decade he began appearing in films and was quite a rarity in that he made the move from silent to sound film, the peak of his career coming with his portrayal of Dr Harding in *Things to Come*. His stage career also kept him busy, and in 1930 Braddell was Noel Coward's understudy in the first theatrical run of *Private Lives*, part of a cast that also included Gertrude Lawrence and Laurence Olivier.

Three years later Braddell and his wife attended a rally organised by Oswald Mosley and were impressed enough to join the British Union of Fascists. So ardent was Braddell's fascism that he became the party's prospective parliamentary candidate for Streatham in 1937 and a friend of William Joyce, later Lord Haw-Haw. Before he could stand for election, however, he was warned that his association with the far right could spell the end of his acting career, so he left the party.

After the war, which he spent as a member of an anti-aircraft battery and then of ENSA, touring Africa and the Middle East, his erstwhile fascism still made acting work hard to come by. Then one day in London's National Gallery he fell into conversation with an Austrian artist turned art restorer called Sebastian Isepp. The pair developed a friendship and Isepp offered an introduction to his trade, at which Braddell proved a natural.

In 1958 he moved to New York and opened a workshop on East 6th Street, where he entered the orbit of Andy Warhol, who cast him in his 1968 film *Flesh*. By the 1970s he was hard up and in poor health, prompting a move back to Kent. He died in 1990 and is buried on a Folkestone hill overlooking the Channel.

When on Wednesday 4 December 1996 fifty-eight-year-old Mrs Marvel Crumpacker of Fort Wayne, Indiana, boarded the Eurostar at Ashford with her daughter Denise Bouwers for a day trip to Paris, she never expected to be descended upon by reporters. Just over two weeks earlier, on 18 November, a fire had broken out on an HGV shuttle train travelling from France to England, which came to a halt almost halfway across. The thirty-one passengers and two crew were evacuated via the service tunnel, some requiring treatment for smoke inhalation. The fire damaged a 500-metre stretch of the tunnel, and services were suspended until repairs could be carried out.

The first Eurostar train to run after the fire left London Waterloo at 5a.m., carrying fifteen crew, ten journalists – and no paying passengers. Services had only been cleared to restart the previous evening, leaving little chance to book onto a train that would normally expect to carry around 200 people. When the train stopped at Ashford Mrs Crumpacker and Ms Bouwers boarded, to find a crowd of reporters surging towards them, just behind Eurostar staff toting a bottle of champagne.

'I was surprised when we arrived at Ashford station and it was so empty; we thought it would be very busy,' said Mrs Crumpacker. 'We are going home tomorrow and this was our last chance to go to Paris. We didn't realise this was the first train.'

'We don't feel nervous at all,' added Ms Bouwers. 'We've heard a lot about Eurostar in the States and I saw it in the film *Mission Impossible.*'

Greenwich-born William Hoskins was a Thames waterman like his father and grandfather before him. In December 1862 the thirty year old had been married for eighteen months and had earlier that year become a father for the first time. On the stormy night of 20 December he was on a boat close to the Maplin Sands off Foulness at the mouth of the Thames estuary, keeping a watch for shipping that might run into trouble in the conditions. The wind became so strong, his boat slipped its anchor and began to drift to the south-east, a movement Hoskins was powerless to stop until he bumped up against a barge called *Matilda* at anchor with her colours at half-mast. On boarding, he found the *Matilda* completely abandoned, and before he could return to his own vessel she too slipped her anchor and began to move with the storm south into the Channel. For nearly two days Hoskins was adrift in foul weather and at the mercy day and night of Channel shipping. Eventually the *Matilda* struck rocks off the French coast and broke up, an exhausted Hoskins washing ashore near Sangatte, a few miles west of Calais, on a bale of straw. The British consul in Calais took care of him until, anxious to let his wife know he was safe, he was well enough to return home aboard the mail packet *Queen*.

On 21 March 1938 Captain Mitchell of the Glasgow-registered steamer *Meta* was passing through the English Channel south of the Isle of Wight en route from Sunderland to Algiers when, as he stood on the bridge, his attention was caught by a distant flash of colour. When he steamed closer he saw a small fishing boat, its sail only half hoisted, on which a man was frantically waving a scarf. When the *Meta* pulled alongside, the crew looked down and saw a young couple, exhausted and frightened, looking back at them pleadingly. They were Franz Nurvinski, twenty-two, and his wife Anna. When they had managed with some difficulty, due to their extreme fatigue, to climb the ladder lowered to them, they were given food and warm clothing. Neither could speak English and explain where they had come from, so Captain Mitchell decided to call at Brixham and put them ashore. There an interpreter was found and the Nurvinskis sat in the mess room of the *Meta* and told British officials their story.

Franz was a German Jew who, having watched with horror the virulent anti-Semitism sweeping his country, had fled with Anna across the border into Austria. All but penniless, they made their way across Switzerland into France with the vague intention of reaching friends in England. Living on their wits, the couple crossed France, heading north until they reached Cherbourg where, as evening fell and they stood on the beach looking out into the wide expanse of the English Channel, they realised they had reached the geographical limit of a European continent on the brink of terrible conflict. With nowhere to stay, the Nurvinskis found an old single-masted boat and dragged it out of sight, closer to the waterline. They climbed inside, pulled a tarpaulin over themselves and, intending to find a way to cross the Channel the following day, fell into a long, exhausted sleep.

When they woke hours later and pulled back the tarpaulin they found the tide had lifted the boat from the beach and sent it drifting out into the open water. It was first light and the Nurvinskis jumped up, looked around and saw water in all directions. They had no idea where they were or how far they had drifted. Franz tried to hoist the lug sail while Anna grabbed the tiller, but neither had a clue what they were doing or in which direction they should aim. When the *Meta* steamed into view Franz grabbed his wife's scarf and began waving it frantically until, to his delighted relief, he saw the ship change course and turn towards them. If his watch hadn't spotted them, Captain Mitchell told the couple, they would most likely have drifted right out into the Atlantic.

British immigration officials listened to the Nurvinskis' story, and contacted the Foreign Office in London and the German consulate in Plymouth, after which it was decided to refuse them entry to Britain. Their interviewers filled out a couple of forms, placed them in their briefcases, stood up, nodded at Captain Mitchell and left, leaving the Nurvinskis with their heads on the table, weeping.

The *Meta* continued with its two extra passengers to Algiers, where the couple were taken into French custody. When he was put in a cell separate from his wife and told he wouldn't see her for the duration of their incarceration Franz was frantic with anxiety and, convinced he would never see Anna again and be deported back to Germany and terrible consequences, tried to hang himself. Prison guards found him in time and saved his life. It was agreed that the Nurvinskis, traumatised by their journey through Europe only to find themselves on another continent altogether, could see each other for a few minutes three times a week.

And that's where the story goes cold, and the fate of the itinerant Nurvinskis is cast adrift like their boat in the Channel.

In 1944 a fourteen-year-old girl called Audrey from Uxbridge in Middlesex was on holiday with her parents in Ventnor on the Isle of Wight. She wrote a letter with a few details about herself, sealed it inside a bottle, and threw it into the English Channel. She'd written half of the letter in English and half in French, in case it reached the Normandy shore, but in fact the letter travelled no further than Portsmouth, where it was picked up by two women. One of the women passed the letter on to her sixteen-year-old nephew Pat Robinson, who wrote to Audrey, and the two youngsters struck up an occasional correspondence until Pat had reason to travel to London as a member of a Royal Marines marching band. They married in 1951 and two years later moved to Canada. On Valentine's Day 2019 the sheltered housing complex in Guelph, Ontario, where the couple now live, posted their story online, leading to worldwide news coverage.

In September 1932 Charles 'Zimmy' Zibelman, a thirty-seven-year-old man from Chicago, made an attempt to swim the English Channel. He set off from beneath the South Foreland lighthouse just north of Dover and initially made good progress. The weather took a turn for the worse, however, and whipped up the sea until after eighteen hours in the water Zibelman was pulled into his support boat exhausted. He'd swum for an estimated thirty miles, but because of the tides he was still a good twelve miles from the French coast. As failed attempts go, his had been a heroic one, but what really made it stand out from those of the other valiant swimmers who'd had to abandon their crossings was the fact that Charlie Zibelman had no legs.

Nobody knows for sure where Zibelman spent his earliest years – he was born in 1893 either in St Petersburg or Chicago to Yiddish-speaking parents – but by the time he was old enough to work as a newsboy the family was settled in Chicago. Charlie was nine years old when he fell under a tramcar on West Madison Street and lost both his legs just below the hips.

Two years later, in 1904, the *Chicago Tribune* reported on an outing for disabled children from the city's tenements that 'Charles Zibelman, a little legless boy' found thrilling because he had never been on a train before. By Christmas 1909 Charlie had more than come to terms with the loss of his legs, giving a demonstration of acrobatics to the boys at a home for juvenile delinquents, telling them, 'Don't chew, don't smoke, don't steal. Be good. Mind your elders and stay away from bad company.' It's advice he might not have taken to heart, as the census the following year lists his place of residence as the Cook County Jail.

By the 1920s Zibelman was an accomplished swimmer, something he attributed to being tipped accidentally from his wheelchair into a pond and finding he couldn't sink. He capitalised on this natural buoyancy by learning to swim and soon proved a popular sideshow attraction. Hawaii proved to be a rewarding location for him: in 1926 he was in Honolulu as part of a troupe of entertainers, and in 1931, the year before his Channel attempt, Zibelman swam for 100 hours in a Honolulu pool; the same year he broke the American record for holding one's breath under water – four minutes, seventeen seconds.

When he arrived in England in the summer of 1932 he was so confident of success that he announced he was going to attempt to swim the Channel in both directions, envisaging being in the water for some thirty hours. After that unsuccessful attempt he returned the following summer, but after ten hours was forced out of the water again by a combination of bad weather and a jellyfish sting on his mouth.

While he never attempted the Channel again Zibelman still clocked up some extraordinary feats. In September 1937 he swam 145 miles in 148 hours along the Hudson River from Albany to New York, smoking 300 cigars, losing 30 pounds in weight and napping in the water – thanks to his remark-able buoyancy. The following year he tried and failed to swim from Miami to Havana, giving displays in Florida during his preparations that included demonstrations of eating, drinking and smoking under water. His last significant feat came in the spring of 1941, when at the age of forty-eight he pulled off a seven-day continuous swim in Honolulu. He spent his retirement in Norfolk, Virginia, where he opened a delica-tessen, a beer house and a string of hot dog stands. He died in 1952.

It would be 1990 before the Channel was conquered by a person without legs, when a Polish swimmer named Lucy Kra-jewska successfully made the crossing.

On 28 September 1926 the body of twenty-six-year-old Luiz Rodriguez de Lara washed ashore at Wimereux on the Channel

coast. A few days earlier he had been seen by two men stripping off his clothes and entering the water at Cap Gris-Nez. A parcel of clothes was found close to this spot where he was last seen, with a note attached that read, 'I have gone to try and cross the Channel without escort. Kindly return my clothes to the Excelsior Restaurant, Boulogne, where I work.' Inquiries at the restaurant revealed that little was known about de Lara or his background, which remain a mystery to this day.

On 20 November 1903, in bad weather and a rough sea, Captain Vampouille of the Dover–Calais mail packet *Pas-de-Calais* looked on from the bridge in horror as a young woman of around twenty years, fashionably dressed in a long dark-grey coat and black hat, walked up to the forward starboard bulwark, climbed onto it and dropped calmly over the side into the sea.

Immediate inquiries produced very little. The woman had bought a through ticket to Paris and had 'spoken cheerfully' to the ticket inspector as she boarded, although she arrived late, moments before the gangplank was pulled up. She had asked a crew member for some ink and paper but none arrived. She left a small string bag in the second-class passenger lounge but it contained no clues as to her identity, just an empty box with the maker's name, 'Dick, bootmaker, Dover', inside. Mr Dick later confirmed that a young woman had bought the box shortly before the packet's departure.

Rumours began to circulate that the woman was sixteen-year-old Carmen Rincke, the daughter of a noted French actress. Three months earlier Rincke, who was working at a Paris convent, had met in the street a young man named Charles de Collanges who immediately made 'burning declarations of love'.

The pair set up home in Collanges's stylish apartment in the French capital, where he promised marriage, then abandoned her penniless in London. It was thought that as a result she had 'determined to end her unhappy existence' on the cross-Channel packet.

Instead, the woman was identified as twenty-three-year-old Helena Rowson Taylor, the daughter of a Liverpool cotton broker, who had been staying with friends in Kensington. She had been suffering terribly from insomnia.

'A little English seaside town, as ridiculous as these sorts of places always seem to be, with too many draughts and too much music.' That was Claude Debussy's verdict on Eastbourne in 1905. He was holed up in the town's luxurious seafront Grand Hotel with a room overlooking the Channel, where he had hightailed it along with his pregnant lover Emma Bardac, hoping to lie low for the summer. Come the holiday season, the bandstand on the promenade was staging daily concerts of the kind of brass-band military *oompah* that would drive a master of impressionism renowned for pushing rhythm, form and structure to its limits absolutely bananas.

For nearly two years he'd been wrestling with an orchestral piece in three movements, and that seafront hotel room a stone's throw from the Channel proved to be the final inspiration he needed to complete one of his most famous works, if not his masterpiece: *La Mer.*

There are few pieces of music, from songs to symphonies, that evoke the sea like *La Mer*, capturing all its moods, rhythms, mystery, violence and serenity in the space of twenty minutes. In the final movement in particular, 'A Dialogue Between the Wind and the Sea', you can definitely hear the English Channel.

He'd begun work on *La Mer* at his in-laws' house in Burgundy in 1903 and finished it in Eastbourne, having left his wife and decamped with his lover.

As a boy Debussy had spent summers by the sea at Cannes and was known to be an admirer of Turner's seascapes, but *La Mer* is unquestionably imbued with the Channel: with Emma the previous year he had worked on the piece first in Jersey, then at Pourville, a coastal village just west of Dieppe. It wasn't rapturously received at first. The French magazine *La Revue musicale* reported of the work's 1905 Paris premiere that the audience 'had expected the ocean, something big, something colossal, but were served instead with some agitated water in a saucer'.

La Mer was playing in my head as I sloshed out of that agitated water and ran up the edge of the saucer to my robe, heavy legs of renewed gravity making me lumber, flinging the garment around my shoulders and wrestling with gloved fingers to yank the zip up to my neck. I pulled my woolly hat down over my head and felt the familiar descent of that first mouthful of tea as it began warming me from the inside, then watched just for a moment the sun easing away from the horizon. I couldn't stay long, though; I needed to return to the warm. As I emerged onto the footpath from the snowy shingle a man in a beanie hat and padded anorak walking his dog looked at me astonished.

'Have ... have you been *swimming*?' he asked through a cloud of breath.

'Er,' I replied as if I wasn't quite sure myself, 'yeah, yeah I have.'

'Blimey,' he said, and looked momentarily startled, as if something had just occurred to him.

'What about your . . .' his brow furrowed as he groped for the most appropriate terminology to use with a man he'd only just met. 'I mean how are your . . . your . . . your crown jewels?'

I'd have to listen to it properly again, but I'm pretty sure that was one aspect of the sea Debussy left out of *La Mer.*

11

Dieppe

The weather at Newhaven was ominous. The early morning shipping forecast intoned gales with high Beaufort numbers and as I sat at the port waiting to board, the wind threw scatters of raindrops at the windscreen and the car shuddered on its wheels. The sky was dark, the clouds thick, the sea a battleship grey. Clearly I was in for a Channel crossing in the great Dieppe tradition.

'Advice to those who would go to Paris by way of Newhaven and Dieppe – don't,' was the terse verdict of one anonymous traveller's account from 1873. Like me he'd found everything reassuringly calm while still in the harbour, but 'scarcely had we cleared the pier heads when the scene was changed. There was a general and undignified scamper from the deck, and the few who remained on it, as a rule, hid themselves beneath bulk-heads or crouched under tarpaulins.'

Once I'd taken a seat in the lounge I could see those same pier heads; I watched as the flat surface of the harbour waters between them turned into a raging, boiling mess beyond. At the end of the jetty stood a lighthouse, behind which a wave now exploded against the harbour wall and the lighthouse vanished beneath a gigantic bloom of white spray. The ship juddered as it left the quay then eased its way out of the harbour. Through the rain-strafed windows I watched the Channel draw nearer, thrashing and surging, firework bursts of spume appearing above the walls. We passed out of the safety of the harbour and into a tumultuous sea.

My 1873 correspondent had ventured below decks at this point where he found an array of passengers already green about the gills, with the voyage only just under way. In the lounge he found his fellow travellers already 'replenishing the basins with which the steward has thoughtfully supplied them'. When a woman he described as 'high-class' passed out and collapsed to the floor she was brought round by her maid throwing a pail of water in her face.

There were no pails of water ready to slosh at the ailing posh on this vessel, but crew members had been passing through the lounge leaving pairs of sickbags on the tables and window sills, placing them carefully like waiters laying napkins, giving reassuring smiles to those blinking nervously up at them at the prospect of their intended purpose. Once upon a time one of those crew members might well have been Ho Chi Minh, who worked as a pastry chef on the Newhaven–Dieppe ferry in the years following the First World War before he returned to Vietnam in 1923.

There was a slight sense of foreboding in the forward lounge that morning, and as we left the sanctuary of the harbour and entered the Channel at its feistiest some lines of Charlotte Smith's verse popped into my head.

> O'er the dark waves the winds tempestuous howl,
> The screaming sea bird quits the troubl'd sea:
> But the wild gloomy scene has charms for me,
> And suits the mournful temper of my soul.

We arrived at Dieppe with sick bags unused, but at the end of a journey that had featured a rivet-rattling battering from the waves and no shortage of collective 'Oooohs' from the lounge when the bows smashed into another breaker to send a mighty skoosh of water onto the windows.

My first proper view of the town was at twilight. Dark clouds still massed overhead, punctuated here and there with the pale yellow of a wintry evening sun, but what really caught my attention was the beach. From the promenade it begins with a bank of shingle, stones of every shade from mid-grey to off-white, all different hues, yet making the whole a uniform monochrome pastel, different from the blacks, browns and yellows of the beach from which I swam every morning at home. The washed-out nature of those colours had an instantly calming effect and drew the eye to the huge, flat expanse of grey sand that shone with the film left by the withdrawn tide and reflected the sky as far out as the tiny waves sprawling almost apologetically onto the shore. A jetty stretched out to my right, while to my left as I looked up at the castle overlooking the town from a hill at its eastern end, the orange floodlights flicked on as if they'd been waiting for me. Even in the gloaming, as the town's lights switched on pin-sharp behind me, I could tell the light here had a special quality. No wonder Turner, Delacroix, Degas, Monet, Pissarro, Renoir, Braque and Sickert had all been attracted to the place.

Walter Sickert had decamped to France in 1898 after leaving his wife, spending his summers in Dieppe from where he wrote to a friend, 'It is bloody lovely here and fucking cheap.' He shacked up with the queen of the fish market, a woman named Augustine Villain, and had a child with her while 'living in a style somewhere between a country gentleman and a pig'.

Most other visitors to Dieppe have been a little more refined. Certain composers were inspired: Camille Saint-Saëns was a native of Dieppe, while Rossini, Debussy, Liszt and Fauré were regular visitors. The town has always attracted writers too:

Chateaubriand, Flaubert, Turgenev, Dumas, Proust and Woolf. One of the first arrived in 1824, William Hazlitt passing through on his way to Italy and being entirely captivated by the town and its inhabitants: 'Life here glows, or spins carelessly round on its soft axle,' he wrote. 'The same animal spirits that supply a fund of cheerful thoughts break out into all the extravagance of mirth and social glee. The air is a cordial to them, and they drink drams of sunshine ... one thing is evident, and decisive in their favour – they do not insult or point at strangers, but smile on them good-humouredly, and answer them civilly.'

Two years later the *Morning Chronicle* visited the town at a time when an improved boat service from Brighton was making Dieppe more popular for English visitors, many of whom made their homes in the town: and not all of them were, like their equivalents at Calais, skint or on the run. 'Dieppe is at this moment the most delightful place in the world for a perfectly idle man,' gushed the *Chronicle*'s reporter. 'It is the Brighton of France, almost exclusively occupied by idle votaries of pleasure.'

Winston Churchill courted Clementine Hosier in the town. They married in 1908 and continued to spend holidays there, falling victim in 1911 to an early paparazzo when photographs of the couple in their bathing suits appeared in *Tatler*. Churchill, thirty-six at the time and home secretary, was pictured on one page reaching for a towel after a swim, while the couple appeared 'disporting themselves in a double canoe' and 'rocking in the cradle of the deep' on another, the giddy scamps.

Even during my first moments in Dieppe I could feel the refinement in the air and already see why it was so popular with the creative. There's space there, the town isn't hunched up against the shoreline as if keeping a constant close eye on the Channel. There's a vast, wide, grassy promenade on the front asserting a firm distinction between the town and its sea. There's an overwhelming delicacy to the light, even with dark clouds massing menacingly out at sea above the pinprick lights of fishing boats turning for home. The English Channel at Dieppe is at its most expansive. There's no sense here of there being an island across the horizon, even though people have departed and arrived here regularly since the second half of the eighteenth century. From here the Channel could be an ocean, there's no sense of the constriction of the Dover Straits – just a feeling of immense space and possibility.

Dieppe, the name coming from an old Norse word for 'deep', was a long-established fishing town until the end of the Napoleonic Wars when it became a destination of choice for the great and the good, largely thanks to the influence of Marie-Caroline, the Duchess of Berry. An Italian noble by birth, Marie-Caroline married Charles, Duke of Berry, nephew of Louis XVIII of France and heir apparent to the throne, in 1816, only for him to be murdered – in front of his pregnant wife – by a Bonapartist assassin when leaving the opera in Paris in 1820. In her widowhood Marie-Caroline took a shine to Dieppe and spent more and more time there, becoming the town's unofficial patron largely responsible for its development as a popular coastal destination. The duchess was to Dieppe what George IV had been to Brighton, attending concerts and plays and bathing in the sea every day during the summer. She made herself visible and accessible, attracting the great and the good from across France and in increasing numbers from the other side of the Channel. Rossini visited in August 1827 in the hope of a musical commission, but declared himself 'horribly bored' and frankly baffled by the duchess's inexplicable delight in dunking herself daily in the sea. Despite this he wrote a short cantata to mark the unveiling of her portrait in the town hall.

Dieppe's location almost exactly midway between Paris and London made it ideally situated as a meeting place for aesthetes, writers, composers and artists, particularly with such an advocate of the arts as the Duchess of Berry in residence for a large part of the year. This attracted attention on the other side of the Channel too, where the British upper classes could never resist a bit of French style, even when their respective nations had recently been trying to kill each other. Before long, particularly when a new service from Newhaven replaced Brighton as the departure point for Dieppe, and cut the crossing time to nine hours, in 1824, an English community began to develop in the town. Indeed, so many fops and dandies arrived from over the water to simper at the duchess that the town soon had a dedicated *quartier Anglais*. During summer months at Dieppe in the first half of the nineteenth century, the number of English visitors and residents swelled to 12,000.

There were ups and downs, not least when the French were overrun during the 1870 Franco-Prussian War and the town filled with 8,000 Prussian soldiers, who arrived prepared for a long siege but instead found little more than a few old veterans bearing ancient muskets and a resigned expression. But by the 1890s it was a fashionable destination once again. Lord Salisbury, three times British prime minister in the late Victorian age, had first holidayed there during the 1860s and liked it so much he bought a holiday home just outside the town. Thanks to Dieppe's reliable telegraph connection with Beachy Head, he was even able to conduct affairs of state from France. Alas, in 1895 he left in a dudgeon not so much high as rocketing through the stratosphere when a new customs chief raised a quizzical eyebrow at the large shipments of whisky Salisbury would collect in person directly from the ship. Previous *douaniers* had for the good of Anglo-French relations tended to look the other way when these 40 per cent proof imports were unloaded, but when the new customs man took the English statesman to one side with the intention of levying duty and a heavy fine, Salisbury threw a tantrum of such haughty self-righteousness it resulted in his stomping off home, packing up his belongings, slapping a 'For Sale' sign in the window and hightailing it back to Britain, never to return.

Walking along the front towards the jetty I couldn't help but notice the succession of memorials commemorating the Dieppe Raid, a disastrous Allied assault on the town launched from the Channel on 19 August 1942. The plan had been for 5,000 Canadian troops supported by 1,000 British and a handful of Americans to storm the beaches and take the town, hold it for a few hours at least, destroy what fortifications they could, then retreat. It was supposed to be easy, a dummy run for a more substantial invasion plan. With Hitler concentrating resources on the Eastern Front, the Allies were so convinced Dieppe would be poorly defended that one senior Canadian officer told his troops, 'Don't worry men, this'll be a piece of cake.' Another benefit of a short, sharp and decisive sortie would be the morale boost it would provide for both troops and public at a time when the war wasn't going well for the Allies, the Dunkirk evacuations still reasonably fresh in the memory.

As it turned out the raid couldn't have gone more wrong. Dieppe's defences were stout: intelligence reports, working mainly from pre-war holiday snaps, possibly featuring Churchill in a one-piece, had missed two massive gun emplacements placed at either end of the town for just such an incursion. The raid began at 5 a.m. and the retreat was sounded before 11. In the meantime most of the 6,000 soldiers who landed on the beach had been killed or wounded by the vastly superior German artillery and air power, or captured and marched through the streets with their hands in the air, destined to see out the war in prison camps.

It was hard to imagine, looking out at the beach and the calm sea beyond, that this had been the scene of utter carnage, bullets zinging through the air from machine-gun posts and from the Luftwaffe above, the beaches littered with bodies, dead men floating in the water, their boots and faces breaking the surface. I read the inscriptions to the sound of the distant gentle lapping of the sea and found it almost impossible to associate this place with the hellish screams of aircraft, the rattle of machine-guns, shells exploding and sending spumes of shingle and dead men into the air, the cries of the wounded and dying – or to believe that this place, so gentle, so steeped in pleasure, could have staged something so dreadful.

I finished my twilight exploration of the front with a walk up the jetty. The night fishermen were arriving in their wet-weather gear, firing up lanterns, flipping open the lids of bait boxes and laying out their rods, settling in for long hours ahead. The fishing boats I'd seen from the beach were chugging into the harbour as I reached the lighthouse at the end of the jetty, where I rested my elbows on the rail and looked out at the darkened sea, watching lightning flashes illuminate the clouds from deep within.

The next morning before breakfast, the sea felt like swimming through silk. The tide was in so there was no repeat of the Calais embarrassment: it was as calm as a swimming pool. The chill of early December was quite bearable beneath a thin, high layer of cloud, the sunlight filtering through and turning the water green. I was the only swimmer. I could see a couple of distant figures strolling along the front, and I could tell from the movement of heads and the fencing of rods on the jetty that the day fishers were taking over from the nightshift. A few hundred yards beyond the jetty I could make out the bright yellow-and-white hull of the Newhaven ferry with its prow facing along the promenade. As I stroked along parallel with the shore, dipping under the surface, the water unfeasibly clear – more like a freshwater lake than the English Channel – I could see why the Duchess of Berry, for all Rossini's bafflement, was drawn into the sea here.

I spent most of the morning on the Grande Rue, Dieppe's main street that runs parallel with the shore, built like the streets of most coastal towns to be protected from the worst of the storms battering in from the sea. I watched the street come to life, awnings extended over shops and trays of pastries placed in the windows of the patisseries. At the centre of the town's main square, which had a condom-vending machine at its corner, was a statue of a raffish-looking man in seventeenth-century thigh-length boots and a rakish wide-brimmed hat – Abraham Duquesne, one of the great French figures of the Channel.

Duquesne was a Huguenot born in Dieppe in 1610, the son of a naval captain who followed his father to sea. He became a captain in the French navy at twenty-five, and when his father was killed in a sea battle with the Spanish he developed a deep hatred of France's biggest foe that wasn't England. He went after the Spanish with some success in a series of battles, until in 1644 he fell out with the French naval administration and went into service with Queen Christina of Sweden, nibbling away at the Danes for three years before returning to service with France when his old nemesis had been usurped. He saw long and distinguished service in encounters with the Dutch and his old foes the Spanish, enough to see him promoted to admiral and handed the keys to his very own castle.

One of his last engagements was the bombardment of Algiers in 1683, in an attempt to liberate French slaves there. The locals were not exactly delighted about this, not least because Duquesne had done the same thing the previous year, and they responded by loading their biggest cannon with shrapnel together with the French consul Jean Le Vacher, whom they stuck head first down the barrel and fired at Duquesne's ship. They enjoyed this so much, they corralled a number of other leading French residents of the Algerian capital as fleshy ammo, and launched them too towards Duquesne's vessel.

Perhaps not surprisingly, the admiral retired the following year, citing ill health, and died in Paris in 1688, but to the people of Dieppe he remains a hero and a symbol of the town's long maritime heritage.

At the western end of the Grande Rue is the old harbour, now a marina but until the 1990s the spot where the Newhaven ferry would arrive and depart, depositing its passengers right in the middle of town. Yachts bobbed at its jetties, and the sun emerged from behind its shifting layer of cloud to sparkle on the water among the pontoons. A century or so earlier this would have been a very different scene, with churning packet ships lining up their paddle wheels against the quay where steam trains phumph-phumphed, and exchanged passengers bound for England for those bound for Paris. The shouts of porters and boat crews would have been lost in the tumult where today the only noise is the campanology of halyards.

The harbour certainly looked a little different at 4.30 in the morning on 20 May 1897 when *La Tamise* arrived from Newhaven and docked, as two men, the Canadian art critic Robert Ross and the English author Reggie Turner, waited outside the customs shed for a familiar figure. They'd already spotted him descending the gangplank, taller than the other travellers, thinner than they remembered, a large felt hat on his bowed head keeping him unobserved. Eventually he emerged onto the cobbles, greeting them in the lamplight with a slightly reticent smile of relief at the sight of friendly faces.

Oscar Wilde had been released from Reading Gaol two days earlier after completing his two-year sentence for gross indecency, and had come to Dieppe to wait out the inevitable brouhaha. He had happy memories of previous visits, once with Lord Alfred Douglas not long before his arrest and once on honeymoon with his wife Constance in 1884. Eschewing the more popular hotels at the quayside, having handed a bulky envelope containing the manuscript of *De Profundis* to Ross, Wilde checked into a small *pension* tucked away in the shadow of the castle at the western end of town, under the alias he'd chosen for the duration of his stay – Sebastian Melmoth. The Christian name was that of his favourite saint; the surname was taken from the title of a Gothic novel, *Melmoth the Wanderer*, written by his great-uncle the Irish clergyman Charles Robert Maturin.

His friends had already filled the room with books and flowers. 'I feel as if I've been raised from the dead,' he said.

Wilde hoped to stay out of circulation in Dieppe at least until the heat was off at home. It seemed like the perfect match: as the English poet Arthur Symons put it, in Dieppe 'life, if you will but abandon yourself to the natural current of things, passes in a dream'. However, a combination of his striking appearance, innate flamboyance and the gossip of a tightly knit English community meant it wasn't long before 'Sebastian Melmoth' was recognised for who he really was. His early days in Dieppe had been relatively carefree – he'd even been persuaded by the English writer Ernest Dowson to visit a brothel and had compared the experience to 'chewing mutton', asking Dowson to 'tell it in England, where it will entirely restore my reputation'. But as word of his identity spread, Wilde found himself increasingly ostracised by the town. Dieppe worthies would cut him dead in the street, café owners fearing the effect on their businesses suddenly found all their tables reserved when Wilde appeared: at one establishment when he arrived in a party of four he was told that alas there were only three seats available.

One of his favourite establishments was the Café des Tribunaux, still there today and still arguably the beating heart of Dieppe's town centre. Situated on a small piazza that opens out from the Grande Rue, it's one of those buildings that doesn't appear to fit the street plan, one the town has developed around. Its striking frontage is a Dutch-style gable dating from the eighteenth century, pre-dating Dieppe's Age of Swank, and its name reflecting its early days as a meeting place for lawyers. Inside, however, you are immediately transported back to the 1890s. Art Deco lamps featuring women in flowing Greek-style robes and holding aloft glass balls of light top the dark wood partitions, while the centrepiece is the huge chandelier suspended from the galleried ceiling. There's a small morning crowd being waited on by attentive staff in white shirts, one of whom shows me to a table and somehow understands my mangled attempt to order a coffee. The only concessions to modernity are two unobtrusive televisions silently running a news channel, and the piped music, just audible enough to mask the specifics of conversation. I can immediately see why Wilde would have liked it here, even if a boisterous gathering of friends visiting from England one night brought a stern warning that any repeat would see the town prefect getting involved – and Wilde's expulsion not just from the café, not just from Dieppe, but from France.

Walter Sickert took great pains to avoid him, while his former friend Aubrey Beardsley, the 'decadent' artist who had designed the posters for Wilde's *Salomé* and who spent his summers at Dieppe, was so perturbed by his presence he packed his traveller's trunk and moved along the coast to Boulogne.

He wasn't without friends in the town: the novelist John Strange Winter, the pen-name of Henrietta Stannard, who was a well-known and well-liked figure in Dieppe, once saw him being loftily ignored in the street by some English visitors and immediately bustled across the road, took him by the arm and demanded loudly, 'Oscar, take me to tea.' The Norwegian painter Frits Thaulow became a close friend after witnessing Wilde being turned away from a café: 'Mr Wilde,' he called out, 'my wife and I would be honoured to have you dine with us *en famille* tonight.'

I lifted a glass of wine to Wilde, then wandered down the cobbled Grande Rue in search of the Café Suisse, situated on the corner looking out across the harbour where 'Sebastian Melmoth' had disembarked and where he would sit at a table beneath its arcade observing the town's busiest spot. The Suisse today is far from the opulent *belle époque* establishment it used to be, a modern revamp without any hint of its previous character. The Suisse was also where the artist Jacques-Émile Blanche, walking one day along the Grande Rue with Sickert, pretended not to see Wilde sitting outside and eagerly beckoning him over – something Blanche later regretted. 'I know for a fact he was wounded to the quick by my action and recollection of the episode still fills me with remorse,' he wrote after Wilde's death.

Such incidents left Wilde feeling both shunned and closely watched in Dieppe, not to mention being aware that the town's telegraph office could distribute across London any perceived scandals within minutes. Hence he decamped five miles east along the coast to Berneval-sur-Mer, staying first in a hotel and then renting a cottage from where he swam in the sea every morning and commenced work on *The Ballad of Reading Gaol*. He also attended mass, despite not being baptised a Catholic. Just before his release from prison he'd written to the Society of Jesus asking if he might be accepted on a six-month retreat once he'd regained his liberty, a request that was immediately denied, almost by return of post.

Wilde's Berneval cover was blown when he invited a dozen boys from the local school to a tea party in July 1897 to celebrate Queen Victoria's jubilee. When in August the Channel weather began to deteriorate, he declared Dieppe 'too British' and headed for Rouen and a reunion with Douglas. The pair then decamped for Naples where they ran out of money, then three years after his arrival in Dieppe Wilde died in Paris.

On my last morning in the town I went to take a last swim, especially now I knew that Wilde had taken to the same waters a little further along the coast. As I mashed over the brow of the shingle I saw in front of me a middle-aged woman in a cagoule, with a rucksack at her feet and a book in her hands from which she appeared to be addressing the sea. The wind had got up and the breakers were crashing a few feet in front of her so it was hard to make out, but on the breeze I could hear snatches of her song as the wind whipped her hair around her head and tugged at the pages she'd anchored with her thumbs. The last thing I wanted to do was interrupt: this was clearly a personal exchange between her and the Channel and, with nobody else around, I was already intruding enough without stripping down to my shorts and prancing into the waves next to her. I walked back to the hotel.

The return crossing was calmer than the one that had brought me to Dieppe and I spent most of it watching out of the corner of my eye as an English man in a shiny grey shirt and with shiny grey hair did card tricks for the teenage French barman, each one producing a loud exclamation of disbelief, an occasional thump of hand on bar and even at one point an outright pirouette. At the next table a man in a fleece turned coins over in his hands, separating his sterling from his euros. Outside in the gathering dark the white rear floodlights of a Boulogne trawler lit up a patch of sea behind us, rising and falling as it crossed our wake, lighting first sea then sky with the rhythm of the waves and bobbing away into the night, rising and falling, rising and falling.

12

Blanchard the Balloonist

It's always struck me as a bit unfair that Louis Blériot has become the poster guy for cross-Channel aviation rather than Jean-Pierre Blanchard. When Blanchard made the first aerial crossing between England and France in January 1785, it was an occasion loaded with significance. As well as having under his belt the achievement of the crossing itself, Blanchard was the first person to see the Channel from above, to observe the layout of the coastline, to see Dover Castle and the walls of Calais as if they were features on a map.

He also brought France and England closer together than anyone ever before. His crossing took two and half hours, and when you consider that he landed a good dozen miles inland the Channel crossing itself can have barely scraped two hours, unthinkable speed for the time. The coasts of England and France were well fortified against attack from the sea, but it hadn't occurred to either nation that one day it might be possible to launch an offensive from the air. When Blanchard made his crossing it was still barely fifteen months since the first ever manned balloon flight, making his Channel crossing as advanced a piece of aeronautics as it's possible to imagine, especially considering he'd only made five ascents of his own before tackling the Channel. Now that we can fly from Britain to much of Europe well inside Blanchard's flight time, it's difficult to appreciate the effect his journey must have had on Britain's national psyche.

Suddenly the Channel wasn't nearly as significant a fortification as it seemed. Not only that, but the man who had undermined it by flying over it was, of all things, a Frenchman.

Another thing that made Blanchard different from Blériot was that Blanchard was actually quite good at flying, enjoying a long and successful career marked by its absence of significant crashes and injuries. He was also, in the true tradition of notable Channel people, a bit of a character.

Born in 1753 to a handyman in a village outside Rouen, Blanchard developed a keen interest in physics and engineering at a young age, and despite his humble origins decided he was already a world-class innovator and inventor. Obviously hugely intelligent, he was an autodidact who took great pride in telling everyone exactly how brilliant he was. Insufferable as that sounds, the difference between him and your average pompous gobshite was that he had the chops to back up his immense self-regard.

One of his first inventions was a dangerous-sounding rat trap involving a pistol, then in 1769 he came up with a velocipede, a sort of bench-on-wheels-cum-proto-bicycle that one rode a little like the Flintstones drove their cars – a machine on which he was able to travel the thirty-five miles between his home at Petit Andely and Rouen in three hours. So sensational was this breakneck-speed machine that it and its reputation took Blanchard to Paris, where the sixteen year old rode up and down the Champs-Élysées throwing alluring glances at potential investors, a number of whom stumped up to provide the young inventor with the funds he needed to develop further projects. At the start of the 1780s a hydraulic pump showed initial promise, but it failed to realise the kind of readies Blanchard needed to develop it properly. By then, though, he'd learned of the miracle of flight – and it was a miracle he was determined to perform.

In 1782 he announced he would be demonstrating his brand-new invention the *vaisseau volant*, a wooden flying machine that had four flappable wings operated by a lever. The results were, it must be said, mixed. Having trumpeted a public demonstration of his miraculous novelty, a large crowd gathered to see it in action only for a heavy rain shower to cause Blanchard to abort the mooted flight, turning what hundreds of people were expecting to be a thrilling demonstration of how man would conquer the heavens into the spectacle of some soggy wood in the rain.

When, clearly deciding the cocky Blanchard needed taking down a peg or dozen, a newspaper printed a cartoon lampooning his flying machine that never flew, it morphed into a popular pamphlet distributed widely around the French capital. The anonymous author implied among other things that his contraption would be of interest only to the kind of men who visited prostitutes. A bit of a leap even for some of our popular columnists today, but Blanchard wasn't down for long. After all, at least people were talking about him.

Then the Montgolfier brothers emerged with their flying balloons, and Blanchard, piqued at these potential usurpers of his destiny as king of the heavens, decided he needed a slice of that action. Somehow persuading more investors to hand over cash – it was the only way to get him to shut up about his immense abilities, presumably – by February 1784 he'd funded and commissioned a twenty-six-foot-high green-and-yellow silk balloon, from which a boat-shaped gondola was suspended. Convinced that the air above us worked on the same principle as the sea, he had equipped his craft with a rudder, aerial oars and a system of silk wings attached to wooden frames that he claimed would make the craft as navigable as a yacht.

An early and brief experiment away from prying eyes at the home of one of his investors proved successful enough for

Blanchard to attempt another public demonstration, scheduled for 2 March 1784 on the Champ de Mars in the centre of Paris, where the Eiffel Tower stands today. Roads were closed and a large crowd assembled, uttering timely oohs and ahhs as the silk balloon inflated and began to strain at its ropes. Just as he was about to climb in and set off skyward, however, a student named Dupont suddenly hopped into the gondola and announced to Blanchard that he was coming too. When Blanchard protested was he heck as like coming too, Dupont threw a tremendous strop, produced a sword and set about slashing at the balloon and its gondola, causing serious damage and wounding Blanchard's hand in the process.

It's probable that had it not been for his previous aeronautic humiliation, Blanchard would have abandoned the demonstration on the spot, but fearing yet more insinuations about brothels, he opted for the show going on. Up he went, into the sky over Paris. The crowd, having seen the Dupont incident and the damage it had caused to the aircraft, were agog, convinced of imminent and spectacular disaster. Yet Blanchard somehow managed to stay airborne for a full two hours, eventually coming in to land near the porcelain factory at Sèvres six miles or so from where he had departed. Navigation having been rendered all but impossible by the sword-wielding student, he'd floated off in the opposite direction to the one announced, about which the citizens of Paris were a bit sniffy, given that ballooning was a pretty common thing these days. They'd only turned out because they were expecting to see Blanchard drive around up there for a bit, as he'd promised, not just drift off in the direction of some pottery.

When a further demonstration at Rouen failed to attract the kind of extra investment he was after, Blanchard folded up his balloon and headed for London. Trumpeting his presence in the British capital in characteristically blustering style, he was, he declared, the world's finest aeronaut, a claim he supported by listing a number of balloonatic achievements in France, nearly all of which he'd completely made up. He bullshitted a range of minted notables from the Duchess of Devonshire – after whom he was prepared to name his balloon – downwards, including an American called John Jeffries. A Harvard-educated physician from Boston, Jeffries had served in the British military during the American War of Independence, whose outcome made it pertinent for him to leave in a bit of a hurry, first for Nova Scotia and then, in 1779, for London. He opened a surgery in Cavendish Square treating the nobs, and by the time Blanchard hove into his orbit Jeffries was a very wealthy man indeed.

Jeffries had long been interested in meteorology – before the war, he'd been one of the first Americans to record weather data – and Blanchard's balloon antics caught his imagination to the extent that he offered the Frenchman £700 towards a balloon flight across the English Channel as long as he could come too. After a successful trial flight over the Thames in November 1784 the pair travelled to Dover a week before Christmas to prepare for their assault on the Channel, where Blanchard ratcheted up the Blanchard quotient to previously unscaled heights.

Despite Jeffries having coughed up most of the expenses for the expedition, Blanchard had been determined all along that he would make the journey alone. The glory of crossing the Channel was to be his and only his. Once the pair had agreed the details, Blanchard drew up a contract with an extra clause: to wit, Jeffries agreed that should it be necessary in order to save the balloon he would jump out of the gondola. It says a great deal for Blanchard's self-importance that he even suggested it; it says even more about Jeffries' eagerness to proceed that he agreed.

Then, once they had reached Dover and based themselves inside the castle grounds, Blanchard sent Jeffries to town on an errand, bolted the door behind him and refused to open it on his return. A baffled Jeffries had already received a visit from a slightly embarrassed Governor of Dover Castle who had come, he said, in the light of a number of accusations by Blanchard of unreasonable behaviour by his co-pilot. All of it was utter tosh, of course, and it was news to the Governor that Jeffries had not only bailed Blanchard out financially already but was also footing the bill for the entire operation.

That wasn't the worst of it by half for poor old Jeffries. The two men inflated the balloon ready to depart on 6 January and waited for a strong north-easterly wind to die down, meanwhile loading their equipment into the gondola, including a pair of cork lifejackets, a packet of promotional pamphlets Blanchard had prepared for distribution over France, thirty pounds of sand as extra ballast and Jeffries' selection of scientific instruments.

When the wind had dropped enough to start the ascent safely the two men stepped into the gondola. Blanchard loosened a couple of mooring ropes and frowned. A couple more, and he began to look pantomimically concerned. Normally the balloon would be straining to rise at this point, he said, but the gondola seemed disinclined to leave the ground. Declaring they were clearly too heavy, Blanchard informed Jeffries that he was very sorry but the doctor's weight was obviously the difference between success and failure and there was only one thing for it, he'd have to step out of the balloon and let Blanchard go on alone.

Jeffries had already noticed that Blanchard seemed to be filling out his overcoat a little more than usual, and it didn't take long for the American, well used to Blanchard's antics, to deduce there was something fishy going on. First asking and then insisting that the Frenchman open his coat, and possibly aided by a little light groping, he discovered that Blanchard had fashioned himself a belt from which he had hung a series of lead weights.

Faced with such superhuman levels of deviousness most normal people would have packed it in by now, or at the very least turned Blanchard on his head and put him into service as a golf club using a selection of suitably sized rocks as balls. But such was Jeffries' commitment to making the trip, he was even prepared to overlook this latest bout of oompus-boompus and still travel with a man whose whopping sense of entitlement made the balloon itself only the second-most inflated thing in Dover that day.

The next day, 7 January, dawned bright, clear and cold over Dover. To Blanchard's delight the wind had finally turned to the north-west, facilitating the prospect of a successful crossing. The pair spent the morning inflating the balloon and manoeuvring it over the walls and out onto the cliffs, tethering it as close as possible to the edge. Then on the stroke of one o'clock the two men unhitched the mooring ropes, and the green-and-yellow balloon rose into the sky, its occupants waving to the masses who had gathered on the cliffs and down on the beach, to provide a 'most picturesque and beautiful view indeed', according to Jeffries.

While they gained height quickly, despite the favourable wind they didn't seem to be making much progress towards their destination, Calais. If anything, they were heading east. Having noted that the conditions were so clear they could make out Canterbury, Jeffries observed waves breaking over the Goodwin Sands, a clue that the balloon wasn't taking the optimum course towards France. He wasn't over-concerned, however, recording 'enchanting views of England and France being presented to us by the rotary and semicircular motion of the balloon and car', which 'greatly increased the beauty and variety of our situation'.

Once they were halfway across, however, the balloon began to lose altitude. Out went the ballast, arresting the descent for a while, but the closer they drew to the French coast the lower the balloon seemed to be sinking. Over the side went the packets of pamphlets.

'We now had nothing to cast out as ballast in future,' rued Jeffries, 'excepting the wings, apparatus and ornaments of the car, with our cloaths [*sic*] and a few little articles.'

Still the balloon seemed to struggle, Jeffries estimating that by the time the French coast loomed ahead it was barely three-quarters full. Out went their food, the wings had to go, also the oars, their last bottle of brandy and even the cloth lining the inside and outside of the gondola, yet still they were heading straight into the chalk cliffs outside Calais. Jeffries doesn't say as much, but it's possible, nay, probably inevitable, that around this time Blanchard pointed out their contractual agreement and offered to give him a leg up and over the side. Either way, things became desperate enough for the two men, despite the intense January cold, to strip down to their underwear and jettison their clothes. Next, they slipped the cork lifejackets over their heads and prepared for the most drastic action of all in order to try and save themselves: namely, to cut the ropes by which the gondola was attached, and hang from the ropes attached to the balloon itself. Then, just as the situation appeared hopeless, the balloon suddenly started to rise. As the two men stared at the approaching cliffs the aircraft lifted gracefully over them, affording a sudden and magnificent view of the French countryside.

'Nothing can exceed the beautiful appearance of the villages, fields, roads, villas etc. beneath us, after having been just two hours over the sea,' sighed Jeffries, but now they had a different problem: the balloon was still rising. 'From the height we were at now, and from the loss of our cloaths, we were almost

benumbed with cold,' recorded the physician of their increasing altitude. It took several miles for the balloon to begin to
descend again, heading straight for the large forest by the old
town of Guînes. Looking for something – anything – they could
jettison to help them over the treetops, Jeffries had a brainwave.

'Happily it occurred to me almost instantly that we might be
able to provide it from within ourselves, from the recollection
that we drank a great deal at breakfast and not having had any
evacuation,' he wrote. There were two empty leather ballast
bags hanging up in the car. They took one each, and to Jeffries'
delight 'we were able to obtain, I verily believe, up to five or
six pounds of urine', which was then cast over the side of the
basket.

Having arrested the descent to a manageable level the men
were able to avoid a violent collision with the trees, and instead
slowed their landing by grabbing at tree tops until the car could
be lowered into a small clearing. Freezing-cold in their underwear, they did their best to keep warm, until they heard the
sound of approaching hooves: the first of the excited spectators
who had followed their progress on horseback burst through
the trees to find two men, each wearing nothing but underpants
and a cork bra, apparently, enjoying a cuddle.

Slightly comic the piddle-tossing conclusion to the flight may
have been, but Blanchard and Jeffries had pulled off something
remarkable – and not just their trousers – ushering in an entire
new era for their respective countries and the wider world at
large. Man had conquered the Channel by conquering the
heavens, and nothing would be the same again. Nobody had
ever crossed the Channel as fast as Blanchard and Jeffries: it
had in the space of a couple of hours been entirely diminished
as a barrier.

The pair were taken back to Calais for some wild celebrations,
and then on to Paris where they were presented to Louis XVI,

who immediately awarded Blanchard a royal pension that set him up financially for life, while also making sure the men were well provided for in other ways: Jeffries' diary candidly records that on their second night in Paris 'a number of French *dames* entered our apartments and embraced us again and again and chanted us several verses honorary of our aerial voyage' (which is a new name for it).

Louis had the balloon and its car sent to the Church of Notre-Dame in Calais for display; it was later transferred to the town hall where it remained until as late as 1966. Jeffries, meanwhile, stayed a few days in the French capital, probably being embraced again and again by more *dames* and getting in some serious verse-chanting in honour of their aerial voyage until he could chant no more. Also, thrillingly for a man of science, he enjoyed several meetings with Benjamin Franklin and received news from London that he had been granted a Fellowship of the Royal Society in recognition of his achievement.

It was 3 March before he returned to Dover, where, already feeling pangs of nostalgia, he 'visited the cliff and spot of departure on our late aerial voyage to France. The recollection was awfully grand and majestick.'

Blanchard and Jeffries went their separate ways after the flight, the American probably keener than most to see the back of the conniving aeronaut and his weighted belt. Blanchard milked his fame for all it was worth, touring Europe giving demonstrations of ballooning, chalking up the first recorded flights in Belgium, the Netherlands, Germany and Poland, and taking off on a commemorative flight to mark the coronation of the King of Bohemia in 1791. Two years later he made the first balloon flight in America, from Philadelphia to Deptford, New Jersey, a journey witnessed by President George Washington as well as future president Thomas Jefferson.

In February 1808 Blanchard was giving a balloon

demonstration at The Hague when he suffered a heart attack and fell fifty feet to the ground. He died as a result of his injuries a year and a month later, aged fifty-three. His wife Sophie continued his legacy after his death by giving demonstrations of her own, becoming a favourite both of Napoleon – especially after she drew up plans for a potential balloon invasion of England – and, after the restoration of the monarchy, of Louis XVIII. In 1819 she was making a flight at the Tuileries in Paris when fireworks she was throwing from the car ignited the balloon, which caught fire and crash-landed onto the roof of a house. She fell to her death.

Jeffries, meanwhile, never went near a balloon again. He remained living and working in London until 1790 when he returned to Boston, where he lived until his death in 1819 at the age of seventy-five.

Six months after Blanchard and Jeffries made their Channel crossing Jean-François Pilâtre de Rozier, a science teacher from Metz, attempted to cross from the French side of the Channel. He had in November 1783 made the first ever free flight in a hot-air balloon in the company of a marquis, a twenty-five-minute journey over Paris during which the balloon rose to an altitude of 3,000 feet. Prior to that achievement he had been assisting the Montgolfier brothers by sending up balloons with a duck, a cockerel and a sheep as passengers, before volunteering to be the first human passenger for an untethered flight. The king had recommended sending up a couple of convicts, given the dangers involved, but Pilâtre de Rozier succeeded in changing his mind.

On 15 June 1785 he set off from Boulogne at 7 a.m. with one Pierre Romain, but no more than half an hour after the launch the wind changed, forcing the balloon back towards the French coast where it somehow caught fire over Wimereux and fell to earth. Both men were killed instantly, becoming the world's first aviation fatalities. There's a monument on the spot where aviation lost its innocence, a small obelisk by a fork in the road only a couple of hundred yards from the sea. It's well maintained, a flower bed next to it, a gravel path leading up to it and a concrete bench placed nearby, a reminder that for all Blanchard's success in drawing two nations closer together with his conquest of gravity the Channel was still capable of baring its teeth when it needed to.

13

Blériot – a Terrible Pilot

There had been a heavy fog. It had lifted now but there were still remnants of haze, and a stillness in the air as if the day had paused. Walking down a woodland path, I emerged into a clearing above Dover Castle. From far below I could hear a faint clanking from the busy port, underpinned by the barely detectable low throb of ferry engines. The clearing had been landscaped, surrounded by a wooden paling into which the initials of courting couples had been carved. There was a scatter of benches facing what I'd come to see, but first I was looking at the sky, clear and blue now the fog had gone, and unusually for this part of the world there was not a single vapour trail in the patch of the heavens I could see above the treetops.

The birds sang loudly all around me, out of sight in the trees, as if they were reclaiming their territory, where man had first tried to usurp them. Finally I looked down at what I'd come here to see, a concrete silhouette in the grass in the shape of an aeroplane. Between the wings, roughly where the cockpit should be, was an inscription. 'In making the first Channel flight by aeroplane,' it said, 'Louis Blériot landed at this spot on Sunday 25th July 1909'. I was standing at the exact location where the English Channel truly stopped being a barrier.

Louis Blériot was a terrible pilot. Absolutely rank rotten. Hopeless. This might sound a strange thing to say about the man whose name is synonymous with the early days of powered flight – like calling Wordsworth a doggerel merchant or dismissing Einstein as iffy at equations – but it's true. Before his cross-Channel flight that day in 1909 Blériot was regarded at best as a bit of a joke and at worst as a dangerously reckless aviator destined for an early demise in a pile of burning metal and wood. He had no grasp of aeronautics and only a cursory understanding of the science involved in heavier-than-air flying machines, even for those pioneering days when aviation involved quite a bit of guesswork and a sense of ah, sure, it'll probably be OK.

What he did have was an unshakeable faith in himself that defied any definitions of logic and eliminated any form of self-doubt whatsoever. His was the kind of colossal self-regard that permitted him, spotting a young woman lunching with her parents across a restaurant one day in 1900, to decide on the spot that he was going to marry her, then go back afterwards and bribe a waiter for her contact details, then four months later to marry her. Maybe even with her consent.

By then Blériot was a successful businessman manufacturing headlamps for the new-fangled cars that were sweeping the wealthy part of the nation. As well as securing himself a wife, the dawn of the century saw him develop a different kind of passion. Having been utterly transfixed by an early flying machine at a 1900 *exposition*, he threw himself into aircraft development, showing an early enthusiasm for ornithopters, machines that sought to fly like birds by flapping their wings and appeared in montages of early aviation accompanied by jaunty soundtracks of banjos and swanee whistles.

By 1905 Blériot had already walked away from a number of the estimated forty crashes he would endure during his lifetime. But his overall crash tally didn't put off another aviation pioneer Gabriel Voisin from going into partnership with the mustachioed gravity-defier, even when Voisin nearly drowned trying to control a Blériot-designed glider, Blériot II, that flipped over and pitched him into the Seine while the man himself stood filming the accident with his (also new-fangled) cine camera.

By 1907 Blériot was five planes into his career as a committed crasher, trashing the Blériot V monoplane more than once without it actually leaving the ground. When he did finally heave the aircraft into the air one spring day that year, he was so surprised that he overcompensated for the ascent and ended up smashing the plane nose first into the turf. He avoided injury – a considerable stroke of luck given that he'd chosen to place the heavy, flammable engine right behind his seat. For his next design he calculated that if his aircraft had four wings instead of two he'd have twice as much chance of staying in the air. And indeed, a few weeks after the Blériot V crash he finally managed to leave the ground for a significant distance – about a hundred feet, and at a giddy altitude of around seven – in the Blériot VI biplane. Even more significantly, he actually landed rather than just ploughing into the ground. Returning to form, though, on his next flight a couple of weeks later he crashed from a height of around twelve feet when a propeller blade worked loose and sent the plane earthward in the now familiar fashion.

No matter how many splinters he had to pull out of his backside Blériot's determination to become a successful pilot seemed only to be strengthened by gravity's total dominance of their tussle. Having fitted a much larger engine, 50 horsepower, in September the same year, when he managed to reach an altitude of around eighty feet it seemed that finally he was actually flying. As he whooped with delight the new engine chose that moment to cut out, sending the plane into the old nose-first descent towards terra firma. As the machine arced towards the ground Blériot somehow managed to climb out of the cockpit and pull himself towards the tail – impressively quick thinking that helped the aircraft to hit the ground in something approximating a horizontal position. He suffered nothing more than a few scrapes and bruises, but the crash spelt the end of the Blériot VI.

The Blériot VII saw a return to the monoplane, one in which he performed a U-turn in the air without mishap, and with the much-improved Blériot VIII he managed a flight of several miles in October 1908 before the aircraft was destroyed in what the records describe as a 'taxiing accident'. By the next summer he was up to the Blériot XI (IX and X never flew after he fell out with the engine manufacturer) and eyeing up the £1,000 prize offered by the *Daily Mail* for the first pilot to cross the Channel by the end of the year in a heavier-than-air vehicle. Despite never having flown for longer than fifteen minutes in one go, he decided the prize had his name on it.

It's hard to comprehend just how impossible a task flying the Channel seemed in the summer of 1909. The Wright brothers' first flights at Kitty Hawk, covering a few hundred feet at most, had taken place less than six years earlier while the Channel represented a minimum twenty-one-mile crossing, with nowhere to land until you reached the other side. Even with a support vessel it was an incredibly dangerous undertaking: the wind at the mouth of the Channel is notoriously changeable and a light breeze at the coast could be a gale in mid Channel. Ditching in the sea was as likely to prove fatal as not.

Despite the dangers, the lure of being the first to cross the Channel in an aircraft proved irresistible to a small group of adventurers for whom the cash prize was a bonus but the achievement was what they relished. Arthur Seymour of Boulogne was one of the first to declare his entry, announcing in early June that he intended to fly his Voisin biplane from Cap

Gris-Nez to Dungeness, with a relay of twenty speedboats strung out along the route to provide assistance should he end up in the water. But Seymour never made an attempt on the Channel and disappeared back into the mists of history.

The forty-three-year-old Portuguese-born aristocrat Charles Alexandre Maurice Joseph Marie Jules Stanislas Jacques, the Comte de Lambert, was the next competitor to announce his intention to snaffle the prize. Lambert had not even been flying for a year when he threw his goggles into the ring, but one advantage he had was that his instructor was none other than Wilbur Wright. While his assault on the Channel would be unsuccessful, in October 1909 Lambert did succeed in flying over Paris and circling the Eiffel Tower.

The clear favourite, meanwhile, was Hubert Latham, a wealthy Parisian and grandson of a successful English merchant who had settled in France in the mid nineteenth century. Latham had arguably the best and most elegant aircraft of them all – an Antoinette IV designed by Henri Levavasseur. Given its powerful 50-horsepower engine, he unquestionably had the revs to get across the water without too much trouble, and he had Channel previous: in 1905 Latham had travelled with his cousin Jacques Fauré in a hot-air balloon from Crystal Palace to the outskirts of Paris.

After journeying widely through Africa and the Far East, Latham had watched a flying demonstration given by Wilbur Wright at Le Mans in 1908 and decided that flying was just the thing for him. He took his first lesson in February 1909 and had such aptitude that within three months he had set the European record for continuous flight (sixty-seven minutes) and, in a demonstration of just how young aviation was at that point, seven months after he'd sat in an aeroplane for the first time Latham was the chief flying instructor at a new military school at Mourmelon-le-Grand in north-eastern France.

He was the first to actually attempt the Channel crossing, taking off from Cap Blanc-Nez close to Sangatte on 19 July and flying for eight miles until his engine failed and he was forced to ditch in the sea. When his support vessel arrived they found the pilot sitting with his feet up on the fuselage smoking a cigarette. His plane was badly damaged, however, and Latham had to send for a replacement, which arrived just as the weather closed in.

Blériot hadn't even arrived at the coast when Latham made his attempt, rocking up with his Blériot XI only on Wednesday the 21st to find a classic English Channel gale blowing.

Although it was the height of summer the bad weather didn't lift until the evening of the 24th when Latham decided that, weather permitting, he would make a second attempt at crossing the Channel at 7 a.m. the next morning. At 3 a.m. Blériot's support team at his base at Les Baraques, situated between Latham's Sangatte headquarters and Calais, awoke to calm weather. Knowing of the latter's intentions, the crew realised they could get Blériot into the aircraft and into the air before their rival had even started his soft-boiled egg and kippers.

As Latham and his crew slept, Blériot's plane was wheeled out while the pilot paced up and down as best he could, given he was walking with a crutch after a characteristic mishap involving his foot and some hot oil. He made a brief trial flight in the gloaming at 4.15 a.m. to make sure everything was in order, then at 4.41, as soon as the sun appeared on the eastern horizon – the competition rules were specific the flight had to be made in daylight – the Blériot XI thundered across the field, lifted into the air and headed out into the pinky-orange haze of a glorious Channel summer morning.

Two Marconi wireless stations had been set up specially for the competition, one on the roof of the Lord Warden Hotel in Dover, the other at Cap Blanc-Nez, and their exchanges followed Blériot's progress.

'M. Blériot has started. Look out for him,' read the first from the French side, and at 5 a.m. precisely they transmitted, 'Let us know as soon as you see him.'

At 5.10 the Dover station telegraphed, 'The torpedo boat is steaming towards Dover harbour as fast as she can. The people on shore expect her in a quarter of an hour. It is a perfect summer morning, the bright sunshine sparkling on the water. A number of submarines have just passed and there is a great deal of shipping in the Channel.'

Ten minutes later another message went out from Dover. 'The torpedo boat approaching the shore but there is no sign of M. Blériot at present,' it read, before adding with tongue-in-cheek understatement: 'A Frenchman arrived here this morning explaining that he had to see M. Blériot and refusing to take a bed. He said he had an appointment with the aviator at 5 a.m. We are afraid he is a little late.'

Then at 5.31: 'There is a rumour that M. Blériot has landed on a cliff on the other side of Dover Castle.' Ten minutes later: 'A police sergeant reports that he saw Blériot flying over a cliff behind Dover Castle' – confirmed by the final message sent at 5.52 that he was 'safe and sound. He is now having breakfast at the Lord Warden Hotel.'

Blériot had in fact arrived at North Fall Meadow behind the castle at 5.12, a flight time of thirty-seven minutes. It's customary in such circumstances to report that the aircraft 'touched down', but this was Louis Blériot, for whom the grace and elegance of flight would always remain a mystery. 'Touching down' implies some level of subtlety and skill, but having circled the field a couple of times to lose height the Blériot XI pretty much just fell out of the sky onto the ground with a heavy thud and quite possibly an inventive Gallic oath. The undercarriage was a write-off, and the propeller was smashed to pieces, but, after the aviation equivalent of a belly flop, Blériot walked from the wreckage uninjured beyond his already heavily bandaged foot.

The Frenchman who had insisted he had a 5 a.m. appointment with Blériot may have been a journalist from *Le Matin* called Charles Fontaine; certainly, it was he who awaited the Blériot plane at North Fall waving a French tricolour in order to guide the airman in to land at the pre-agreed spot.

The more I found out about Blériot's flight the clearer it became that it was achieved despite the pilot as much as because of him. Blériot didn't so much leave things to chance as wilfully abandon them to it. So many things could have gone wrong, even allowing for his track record of leaving piles of splintered wreckage in fields while walking away chirping, 'Ah well, back to drawing board.'

He wasn't so much unprepared as completely in the dark as to how he might pull it off. Latham had travelled to Dover to familiarise himself with the locality and the terrain, which allowed him, for example, to choose a preferred landing spot from a number of options. Blériot had never even been to England before: all he knew was to look out for a little French fellow waving a flag in a field near a castle. Having assumed even he couldn't get lost on a straight twenty-mile run between two enormous rocks, he hadn't bothered to take a compass with him, something he came to rue when a mist descended halfway across. He lost his bearings completely, the wind carried him off his intended course, and it was only when he caught sight of the cliffs away to his left that he realised he was heading straight out into the North Sea.

Other than Fontaine the only people to see Blériot land were a couple of policemen, the son of a coastguard living at a clifftop cottage nearby, a handful of soldiers out on exercise and, according to one source, a tramp who'd been sleeping under a hedge. One of the policemen ran to the nearest telephone and reported, 'That flying man is here.' So certain had the *Daily Mail* been that it would be Latham appearing out of the morning haze that their reception committee had gathered at his proposed landing spot at nearby Archcliffe Fort, so when they heard the buzz of an engine away to the east they were forced to lurch up the hill past the castle in their spluttering car, praying they hadn't missed their scoop. Not far behind them were two British customs officials with a sheaf of paperwork for Blériot to fill in. After a brief discussion during which the airman pointed out there was no aeroplane option in the section headed 'Type of Vessel', it was agreed to classify the Blériot XI as a yacht.

In joking about Blériot's abilities and preparations, it's easy to overlook the enormous courage it took to make that flight. The Blériot XI was flimsy. You wouldn't try to sail over the Channel in a vessel so insubstantial, let alone do 50mph 300 feet up in the sky. The monoplane was twenty-five feet long with a wingspan two feet shorter, the body basically a coffin-sized box in which the pilot sat with his top half exposed to the elements, wings and fuselage made from strips of ash and poplar strengthened by piano wire. It was a go-cart with wings, a pedal car in the sky.

In that clearing above Dover Castle, at the Blériot memorial, the outline of the aircraft seems absolutely tiny and you think, there's no way something that small could have made it across with a man inside and an engine at the front. Looking down at it in the grass at my feet, it seemed Blériot's moustache would have given the XI as much lift as those wings. Its engine would barely power a sit-on lawnmower today.

In some ways the plane was only half-finished when Blériot crossed the Channel: it had no instruments, the wing-warping system he had developed to help control the machine hadn't been fully tested, and the engine was built not by one of the top aeronautical engineering firms but by an Italian motorcyclist called Alessandro Anzani, a volatile character who spluttered and spat as much as his engine did. Added to that, Blériot was on crutches and, on the morning of the flight, bleary and hungry, which makes his crossing seem even more of a ridiculous achievement. If the aviation gods had been in any way disposed towards Latham – a highly professional, organised pilot with a brilliant support team – Blériot would hardly have got off the ground that morning, let alone become one of the most famous aviators in history.

Of all the potential pilots to be first across the English Channel he had barely been in the also-rans, let alone among the favourites. Entirely self-taught, he became a pilot through a mixture of bloody-minded determination, wilful ignorance of his own limitations, and a combined fearlessness and foolhardiness that rendered him willing to take to the skies in half-tested machines flying almost literally on a wing and a prayer. *The Times* summed him up politely in its obituary, acknowledging his 'courageous impatience'. 'I have no apprehensions, no sensations, *pas du tout*,' said Blériot when asked how he felt, eloquently expressing what so often tends to be the mundanity of true greatness. 'The moment is supreme, yet I surprise myself by feeling no exultation.'

Latham, meanwhile, had prepared his aircraft for a take-off assuming that at that moment his rival was most likely standing, hands on hips, regarding a pile of wreckage in a field somewhere not far away. Then just as strong winds began to gust up, preventing any possible departure that day, the news reached him of Blériot's success. Latham broke down in tears. He'd done everything right, prepared properly and planned meticulously, practised his technique and tested his equipment, assembled the best support team in the business, paid heed to the weather and respected the Channel – only for a limping Frenchman to lick a finger, hold it up to the wind and fly off into the history books.

Latham's response was to push himself and his aircraft to the limits. A month later he set a new world altitude record at Reims, where he topped 500 feet, then at Blackpool in October he flew in winds he should never have attempted, estimating afterwards that at one point the following wind must have reached 100mph. It was almost as if he was compensating for his caution in those July days at Sangatte by taking his aircraft and his body to previously uncharted extremes: a year after that Channel summer, he raised his altitude world record first to 3,600 feet and then to 4,541.

In 1911 he survived two substantial crashes almost completely unscathed, one into a hillside in Los Angeles and the other onto the roof of a garage at Brooklands racetrack in Surrey, increasing the sense that he was tweaking the noses of the fates he'd trusted so erroneously. Latham died in 1912, not in an aircraft cockpit but on colonial business in what was then the French Congo, some reports saying he was mauled by a buffalo on a hunting trip, others that he was murdered by his porters.

Blériot, meanwhile, saw sales at his aviation company soar in the wake of his achievement, as the rich and adventurous ordered aircraft from the world's most famous flyer. He continued flying in exhibitions and races for a year after his Channel crossing, until in December 1910 a crash into the roof of an Istanbul house while giving a display in blustery conditions broke a number of his ribs and left him with severe internal bruising. Sensing that his aviation luck was finally starting to run out, he gave up flying on the spot. In 1927, less than two decades after his Channel flight, he was at Le Bourget airfield on the outskirts of Paris to see Charles Lindbergh land at the end of the first transatlantic flight. If Blériot had closed the gap between two nations, Lindbergh did the same for two continents.

Blériot, who died of a heart attack in 1936 at the age of sixty-four, and Latham were far from the last significant Channel aviators, however. Others followed who, while not attaining the immortality of being first, still deserve recognition not least because many of them were remarkable characters.

John Moisant, for example, is another landmark figure in the Channel story, having on 17 August 1910 become the first person to fly across it carrying passengers. Born in Chicago in 1875, he'd opened a coffee plantation in El Salvador with his two elder brothers who in 1907 found themselves arrested for plotting to overthrow the government of President Fernando Figueroa. When the US government declined to assist, Moisant went to neighbouring Nicaragua and persuaded them to lend him 300 troops and a gunboat. The mini-invasion was daring but a failure, putting a price on Moisant's balding head and condemning the elder Moisants to a death sentence. The latter saw the USA finally intervene and the brothers were eventually released, but John Moisant retained a burning antipathy to Figueroa and, reading about the pioneering aviators of France, resolved to travel to Europe, learn to fly and use aerial superiority to achieve regime change in El Salvador.

He was a fast learner, and when he crossed the Channel as part of a marathon multi-stage flight from Paris to London a year almost to the day after Blériot's, it was only the sixth time he'd ever been at the controls of an aircraft. This inexperience could explain why he decided to set off in the teeth of a strong westerly wind, one that delivered a freezing hailstorm into his face as he crossed. His passenger was his mechanic Albert

Fileux, and when the farmer in whose field they landed, in the Kent village of Tilmanstone, arrived on the scene the two men were stamping their feet and flapping their arms about in an effort to warm up. Their eyes were raw from the constant wind and when the farmer bathed them with water from a cup, Moisant accepted the receptacle as a gift and later had it engraved with his and Fileux's names. It's possible that Moisant also engraved the name of the third passenger on that historic flight: his cat, Mademoiselle Fifi, who travelled everywhere with him.

In December 1911 Moisant decided to have a crack at a $4,000 prize put up by the Michelin company for the longest flight of the year. His was a last-minute decision: it was New Year's Eve when he took off at 10 a.m. in front of a small crowd in a field outside New Orleans, performing a few stunts in preparation for an assault on the prize. As he turned to come into land, however, a gust of wind caught the plane and tipped it, throwing Moisant from the machine and causing him to plummet twenty-five feet to the ground, breaking his neck on impact. According to some sources, this was the first flight on which he hadn't been joined by Mademoiselle Fifi.

Two months before Moisant's crossing, Charles Stewart Rolls had achieved the first there-and-back flight across the Channel. Best known as one founding half of the car and engine manufacturers Rolls-Royce, Rolls had been a keen flyer since the early 1900s when he took up hot-air ballooning, making nearly 200 ascents and in 1903 winning the James Gordon Bennett Medal for the year's longest continuous flight. A committed adventurer, the same year Rolls broke the world land-speed record in Dublin, reaching 93mph in a French Mors car.

It was a meeting with Orville and Wilbur Wright in New York while visiting a motor show in 1906 that led to Rolls transferring his aeronautical allegiance from balloons to aeroplanes. Two years later the brothers took Rolls for a four-minute flight at Le Mans, an experience that persuaded Rolls to learn to fly himself. In October 1909 he took delivery of a Wright Flyer made by the Shorts company and the following spring gained his pilot's licence from the Royal Aero Club, an institution he'd helped to found in 1901 as an organisation for balloonists. Three months later he became the first Englishman to fly across the Channel and the first pilot from anywhere to do it both ways. On 10 June 1910 Rolls took off from the hangar he'd constructed at Swingate near Dover at 6.30 in the evening, arrived over Calais, circled the town for ten minutes or so, dropped a message of greeting wrapped in a tricolour, then set off for home. A crowd of around 3,000 had gathered at the hangar to see Rolls depart, and those who'd stayed to witness his return were rewarded when at eight o'clock his aeroplane appeared out of the evening haze, crossed the coast over Dover harbour, circled the castle and the Blériot monument, acknowledged the ships and tugs sounding their horns and whistles, and came in to land precisely in front of his hangar and the throng.

It must have been mindblowing for those who witnessed it to think that someone could depart from a spot right in front of them, travel all the way to France and then return to exactly the same spot in no more time than it takes to play a football match. Truly the Channel was shrinking, and fast.

Barely five weeks later, Rolls was dead. Competing in a flying event at Bournemouth on 12 July, he circled the arena ready to attempt touching his wheels down inside a chalked circle when the tailpiece of his aircraft suddenly detached with a loud crack. The machine turned on its back and dropped sixty feet to the ground right in front of the packed grandstand. When concerned onlookers ran to the wrecked plane it was too late. Charles Rolls was the first British aviation fatality. His statue stands on the front at Dover, in flying gear, hands behind his back, looking out at the Channel horizon.

You'd be forgiven for thinking that cross-Channel aviation was an all-male closed shop. While it's certainly true that men, notably rich men with a great deal of disposable income, dominated early flight, there were, as in the case of the Channel swimmers, several women significant to this tale who were drawn to both the skies and the stretch of water between Britain and France. Their names will be new to most, but these are women of the Channel whose stories deserve to be heard.

The first woman to fly across the Channel wasn't a pilot and was never a pilot. Eleanor Trehawke Davies never flew an aircraft in her life – she made that pioneering crossing as a passenger. This is nowhere near enough to prevent her being one of the most important figures in the story of British flight, however, and particularly in the story of cross-Channel aviation.

Trehawke Davies was thirty-two on 2 April 1912 when she crossed the Channel sitting behind Gustav Hamel in her two-seater Blériot monoplane. The pair had left Hendon aerodrome in north-west London at 9.30 that morning with plans to be in Hardelot, just along the coast from Boulogne, for lunch and then, if the weather was agreeable, Paris in time for dinner. They passed between the harbour piers of Dover at around 11a.m., flying at 3,000 feet, and headed across the Channel for Cap Gris-Nez, arriving at Hardelot at 1.45 p.m. After a leisurely two-hour lunch to toast Trehawke Davies's status as the first woman to fly across the Channel, the pair climbed back into the plane and were in Paris for 5.30 p.m.

'What I remember most vividly about my first Channel crossing is that I had to work the second pressure pump,'

Trehawke Davies recalled. 'In mid Channel, at a height of 3,000 feet, when we could see neither sea nor sky for fog and were both numb with cold, Mr Hamel suddenly cried, "Pump! Pump, for God's sake!" and I pumped with all the strength I had.'

Trehawke Davies, despite great contemporary fame, was always a bit of an enigma. Although she enjoyed the publicity her aviation exploits attracted, posing nay, hamming it up, willingly for photographers on her aircraft in leather coat and flying cap, little seems to be known about her beyond her aviation exploits. Indeed, when she died after a lengthy illness in November 1915 it was four months before her death was announced, and even then it was only because she had valuable stuff to be auctioned off. All we know for sure is that she was born in August 1880 into a middle-class family, her father a solicitor and freemason, her mother the manager of a millinery, and was sent to a boarding school in Brighton. When her parents died within six months of each other in 1907 she was left a fortune (and she herself left a sum worth more than a million).

She never married and never learned to fly, yet she owned at least two aeroplanes; a visit to Hendon Aerodrome to witness the start of a round-Britain air race in the spring of 1911 had left her spellbound by the flying machines. Having watched the pioneer-pilot Claude Graham-White take Lady Northcliffe up for a spin, Trehawke Davies begged him for a turn. He eventually agreed, thereby firing a passion for aviation in his passenger that led to her flying with some of the greatest pilots in the era of early aviation, most notably Gustav Hamel.

The German-born son of Edward VII's personal physician, Hamel had learned to fly in 1910 at Blériot's flying school near Paris, where the man himself hailed him as one of the most talented pilots he'd ever seen. On receiving his licence in 1911 Hamel competed in a number of races and displays around

Britain, as well as participating in a demonstration to government ministers of the potential of aircraft for military purposes.

Trehawke Davies told a newspaper in 1914 that she had crossed the Channel by air eleven times in all, nine of them with Hamel and the others with Henry 'Otto' Astley.

She passed the time in the air by writing in her journal. On one particularly dangerous flight with Hamel in the spring of 1913, she was convinced their number was up, and recorded: 'This is our last moment alive in the air, it will be our first moment dead on the ground.'

Her closest brush with death was on her last crossing with Astley shortly before his own demise in September 1912, when her aircraft suddenly plummeted 150 feet to the ground on the outskirts of Lille. 'The plane began a series of plunges and swerves,' she recalled, 'and I felt splinters of wood hitting me that must have come from the propeller. The machine veered over until she reached a vertical position, the left wing in the air, the right pointing to the ground.'

Despite fearing she would be thrown out of the machine to her death, Trehawke Davies maintained an impressive calm. As it turned out, the plane's wing struck the ground in a beetroot field and ploughed along it for long enough that when the body of the aircraft hit the ground it was at a speed slow enough for both the machine's occupants to survive. Astley extricated himself from the wreckage to find his passenger on her feet taking photographs of the destroyed plane. By the time locals arrived they were sitting on a broken wing drinking milk and eating biscuits. Four days later Astley was dead, killed in a flying accident in Belfast while Trehawke Davies was in Paris.

'This is dreadful,' she said on hearing the news, 'and yet I feared it would happen. I asked him not to go without me as I am a mascot for fliers.' It was a car accident in December 1913, close to Hendon aerodrome, while travelling with Astley's widow that exacerbated Trehawke Davies's ill health and hastened her death in 1915 at the age of thirty-five. In January 1914 she had defied her doctor's instructions to rest by travelling to Hendon, and not only flying with Hamel but being in the aircraft while he performed a loop-the-loop.

'I was no more aware of being upside down,' she said of the experience, 'than a spoon is of the sensation of taste.'

The accident seemed to affect her mental health as well as making her physically frail: an interview she gave to an American newspaper in February 1914 described her as 'dangerously ill' and found her in bed, where she'd been since her loop-the-loop flight a month earlier. According to friends, she made her will as many as forty times in the twenty-two months between the accident and her death, each time leaving everything to a single different person before tearing it up and making another. Her obituary referred to two bouts of 'special treatment for her nerves' during this time. She'd lost two of her closest aviation friends to accidents, first Astley and then Gustav Hamel. Five months after looping the loop with her on 23 May, Hamel set off from Villacoublay near Paris to return to England for the London Air Derby. He never arrived. Six weeks later a French fishing boat found a body floating in the Channel off Boulogne. It was in too great a state of decomposition to make recovery possible, but the flying clothes and a map of southern England sticking out of a pocket made it all but certain it was Hamel. On the day of her death, 22 November 1915, Trehawke Davies had remarked upon how much better she was feeling before taking a prescribed afternoon nap. When she awoke she asked her maid for a glass of milk, and in the time it took for the maid to turn away, pour the milk and turn back again Trehawke Davies had died from heart failure.

Her instructions were that her death should be kept secret, that nobody should come to her funeral, that she be cremated and her ashes flung to the four winds. 'I hope when death does come I shall fall several thousand feet and be killed instantly rather than drop from a short height and stand the chance of being horribly maimed yet still alive,' she told her 1914 interviewer. Then she revealed why she always preferred to ride shotgun rather than take the controls herself.

'I really am quite an early Victorian, not a suffragette nor a philatelist,' she explained, somehow conflating having the vote with stamp-collecting. 'I've never played golf or tennis in my life and it's because of my old-fashioned ideas that I've contented myself with being a passenger in an aeroplane and never attempted to drive one myself. I don't believe in doing indifferently what a man can do well.

'I fly because I love the sensation,' she continued. 'It is the champagne of motion and it appeals to me particularly because I am introspective by temperament and one is so detached and elemental up there in the clouds.'

One woman who might have bonded with Eleanor Trehawke Davies over their respective love of flying but would have strongly disagreed about where a woman should sit in an aeroplane was Harriet Quimby, one of the great Channel heroes and a woman whose name would probably be much better known were it not for a deeply unfortunate historical coincidence.

In the spring of 1912, less than three years after Blériot's flight, Quimby became the first woman to fly the English Channel. Compared with the rest of her life it was almost a run-of-the-mill achievement. She was a model, a popular journalist and she wrote seven Hollywood screenplays. She was America's first woman pilot, performing stunts and competing in races as a member of John Moisant's flying team, and was renowned for her stylish purple flying outfits. If Blériot had brought pragmatism and brute force to cross-Channel flying, Harriet Quimby brought style and panache.

She was born in 1875 to William Quimby, a former Union Army cook turned unsuccessful Michigan farmer, and his wife Ursula. Harriet was their tenth child but only the second to survive infancy. Her mother was determined that her two surviving daughters would escape the farm work that had beaten her down for her whole adult life, and encouraged them to read and widen their horizons (Helen, Harriet's older sister, eloped at fourteen and was never heard from again). The farm went into receivership when Harriet was ten and Ursula, who helped make ends meet by selling her herbal remedy Quimby's Liver Invigorator and Blood Purifier, persuaded William to move west to California. When William made a hash of that venture, Ursula – the couple were now in their sixties – instigated a move to San Francisco, where William became a door-to-door salesman selling his wife's potions.

Despite the upheavals and relentless poverty Harriet graduated from high school in 1897, and while Ursula tried to usher her towards journalism she was developing ambitions to be an actress. By 1900 she was modelling part-time while working in a shop during the day and at a theatre in the evenings, which is where she met and befriended D.W. Griffith and his wife Linda, who like Harriet were actors struggling to make a name for themselves. She began writing reviews and arts features for the *San Francisco Chronicle* until in 1903 she'd saved enough money to relocate to New York and secure a job at *Leslie's Illustrated Newspaper*, adding a travel-writing string to her journalistic bow. By living frugally and working hard she was able in 1908 to buy a car and become the first woman in the United States to earn a driver's licence. The following year Griffith also arrived in New York, where he ended up running Biograph

Studios and engaged Quimby as a screenwriter, making films from her seven scripts.

In the autumn of 1910 Quimby's friend Matilde Moisant, the sister of cat-loving aviator John, invited her along to the Belmont Park racetrack on Long Island to see the start and finish of the Statue of Liberty Air Race, only the second public aviation event staged in the USA. As the machines buzzed overhead Quimby turned to Matilde and declared, 'That actually looks quite easy. Why, I believe I could do it myself.'

The following evening Harriet arranged over dinner with the Moisant siblings that John would take her on as a pupil the following spring. His death a couple of months later meant that two of the great names of early aviation never got to fly together, but Alfred Moisant, a third sibling, took over John's flying school and on 10 May 1911 Harriet Quimby at the age of thirty-six embarked on a five-week course there, her fame as a journalist and the novelty of her gender necessitating her disguising herself as a man. Her cover was immediately blown by the *New York Times* – 'Woman in Trousers Daring Aviator' it trumpeted, blurring the question of whether the trousers or the aviation were the bigger deal – meaning she was practically forced to begin writing about her aviation adventures for *Leslie's Illustrated Newspaper*.

Quimby gained her pilot's licence at the second attempt in July 1911, the first American woman to pass. When she beat the French flyer Hélène Dutrieu in a race, winning the small matter of $600 in the process, she became a member of the Moisant flying team and toured Mexico in autumn that year, a trip cut short by the Zapata rebellion. By early 1912 Quimby had her sights set on a conquest of the English Channel. She sailed to England on the *Amerika*, carrying a letter of introduction to Blériot. When she arrived in London she made straight for the offices of the *Daily Mirror* and pitched them her Channel ambitions, her journalist's instincts telling her there was money to be made from an exclusive deal with a single newspaper.

From London she sailed across the Channel to meet Blériot at his factory and secure for herself one of his 70-horsepower monoplanes. The aircraft wouldn't be ready for several weeks, however, so she arranged for the loan of another, less powerful, machine for her attempt on the Channel.

Quimby was in France when Eleanor Trehawke Davies and Gustav Hamel made their Channel crossing on 12 April, depriving Quimby of the accolade of being the first woman to cross the water by air. But being the first woman pilot to make the crossing was a far greater laurel, of course. Quimby had based herself at Hardelot, out of the way of prying eyes. This was where Trehawke Davies and Hamel had staged their celebratory lunch, though it seems the trio of Channel pioneers didn't meet. At Hardelot, Quimby found the wind too gusty to even practise on the unfamiliar machine, which contributed to her decision – prompted mainly by the risk of smashing into the White Cliffs – to start her flight from Dover.

Staying at the Lord Warden Hotel as 'Miss Craig', Quimby had her aircraft shipped surreptitiously across the Channel and waited for the weather to improve. The first day on which the blustery conditions subsided to make it perfect for flying was 14 April, a day ruled out by her habit of never flying on a Sunday. The following day saw the wind pick up again, and she and her coterie, which now included a curious Gustav Hamel, spent a cold morning at Dover aerodrome waiting in vain for the anemometer to stop spinning so fast. So despondent were the members of the party that as Quimby warmed herself by the fire at the Lord Warden that evening one of her group, possibly even Hamel, offered to make the trip in her place, squeezed into her trademark purple flying suit to give the impression she was at the controls.

'The suggestion struck me as so ridiculous I burst into laughter,' she said later. 'I fear I might have hurt his feelings for the offer was made out of the kindness of his heart.'

The next morning dawned still and clear, and when Quimby and her team rose at 3.30 they knew it was on. They were at the aerodrome within the hour, and as she prepared herself Hamel took the Blériot monoplane around the airfield to warm the engine. There was no time for the kind of trial flight Blériot had made before his crossing, as Quimby feared the wind could return at any moment. As soon as Hamel pulled up and climbed out, Quimby leapt into the pilot's seat, tied a hot-water bottle to her midriff, pulled down her goggles and thundered across the field. The wheels left the ground, and the plane rose into the air, circled Dover Castle and headed out across the water.

'I had been climbing steadily to around 6,000 feet, thinking to get above the mist that dimmed my goggles,' she recalled afterwards. 'They were shoved up on my forehead as useless and the mist dampened my face and my clothes. I could see little except my engine ahead of me and my wings at my sides. I kept my attention on my compass and thanked the stars for such a wonderful invention.'

Quimby then descended, hoping visibility might be better lower down, during which there was a brief moment of concern when her engine began misfiring, but when she dropped to 1,000 feet she found the engine running perfectly. The aircraft, however, remained engulfed in fog. Descending even further to 300 feet, she looked down and to her surprise found that she had already crossed the coastline. Still flying by compass, something she'd never done before, she flew along the coast hoping to recognise Calais, but it wasn't where she thought it was: the side-wind had blown her further west than she realised.

She flew past Boulogne, then minutes later came down on the beach at her old base at Hardelot.

'Alighting from the plane I jumped down onto the hard sand and for a few moments stood there absolutely alone,' she said. Then the doors of a row of nearby cottages flew open and their inhabitants came running out towards her, whooping with delight. Quimby sat on the sand and wrote some cablegrams in the margin of a newspaper someone produced, and dispatched a local boy to the nearest telegraph office to pass on the news. By then, people that Quimby had met on her previous visit had arrived, and they carried her shoulder-high from the beach to a nearby cottage for a well-earned cup of tea.

'[A] humble but hospitable fisherwoman brought out her choicest china,' she cooed. 'To me fell the honour of using an immense bowl-cup. It must have held at least a pint.'

In some ways that giant cup of tea was about as sumptuous as recognition of Quimby's achievement got, because the timing of her flight made her the victim of one of the most unhappy coincidences of the twentieth century: the previous night the *Titanic* had struck the iceberg and sunk into the icy waters of the North Atlantic. The world's news agencies and newspapers were entirely preoccupied with the era's most spectacular disaster, so Quimby's success, despite her careful media strategising, earned barely a mention. The *New York Times* noted sniffily that crossing the Channel by air 'is now hardly anything more than proof of professional competency', and went on to advise that 'the feminists be somewhat cautious' of celebrating her achievement too hard, as 'it proves ability and capacity but it doesn't prove equality'.

Quimby sailed for the US and arrived in New York on 12 May to find a city still coming to terms with the scale of the *Titanic* disaster. Indeed, almost straight from disembarkation she went to a memorial service for some of the victims. There was no ecstatic welcome for her and certainly no ticker-tape parade up Fifth Avenue; her flight seemed trivial by comparison with the catastrophe at sea.

In July she travelled to Boston for an air meet at Squantum airfield in Massachusetts, where she was to be the celebrity headliner intending to mark the occasion by breaking Graham-White's 58mph speed record over water. She would have a passenger with her for the flight, event organiser William Willard, to verify her speed, and would fly the same course from the airfield to Boston lighthouse and back as Graham-White had flown to set his record the previous year.

Around 1,000 people watched the pair climb into the aircraft – Quimby in her trademark purple flying suit, Willard reversing his cap for streamlining – and start the engine, take off and fly out over the water towards the distant lighthouse at an altitude of around 3,000 feet. After she'd circled the lighthouse she flew back towards the airfield bathed in the orange light of a vivid sunset, then descended until she was around 1,000 feet above the shallows approaching the airfield.

What happened next was all but inexplicable. The plane's nose dipped suddenly and the crowd screamed as they saw Willard fall from his seat and plunge into the water. The sudden change in weight distribution caused the plane's tail end to lift sharply upwards, enough to pitch Quimby out of the cockpit too, to further screams. The pair fell from several hundred feet up – so close to the shore they hit water barely four feet deep, giving them practically no chance of surviving. When the bodies were recovered it was found that Quimby had been killed on impact while Willard had drowned. Both had suffered terrible internal injuries as well as a host of broken bones.

The exact cause of the accident was never established beyond doubt. The most plausible theory was that Willard had stood up and leaned forward to speak to Quimby and lost his balance, disturbing the plane's equilibrium and pitching Quimby into the void after him. It happened so quickly the pair landed practically on top of one another, the plane crashing a few feet away. Ten weeks after she'd crossed the Channel Harriet Quimby, pioneering aviator, was dead.

Coverage of the tragedy was sniffy and snarky as soon as the day after the accident. For one New York paper, 'the sport is not one for which women are physically qualified. As a rule they lack strength, presence of mind and the courage to excel as aviators.' Another cited the 'horde of reckless seekers of notoriety' who craved 'the crowd's applause and the money that comes with that applause'. 'Women like Miss Quimby do little to further the conquest of the air,' said yet another. 'Their fate only serves as another example of the price of human folly.'

Perhaps the most poignant quote came from the handwritten note Quimby had sent to her parents just before departing for Boston. 'If bad luck should befall me,' it said, 'I want you to know that I meet my fate rejoicing.'

There's a photo of Harriet Quimby taken on the morning of her cross-Channel flight. She's standing at the foot of the Blériot monument, where she's stopped on her way from the hotel to the aerodrome at Whitfield from which she's about to fly. She's wrapped up against the cold, an overcoat and sealskin stole over her flying suit and a hat secured to her head with a scarf. She's leaning forward slightly, holding a muffler and reading the inscription, the expression on her face a mixture of awe and excitement, her mouth slightly open, as if she's talking to the monument – as if by being on the spot where Blériot landed she's just been struck by the enormity of his achievement and the enormity of the task that lies ahead of her. It's an unforgettable image. The monument is still bright and new and has already received so many visitors that the ground around it is as worn and shiny as the day the plane landed and practically the whole of Dover came to look at it – when the aircraft was removed the crowds had stood so close, there was the perfect shape of it in virgin grass among the mud, making the siting of the monument a piece of cake. There are no crowds in this picture, though; it's a moment of solitude at first light, a communication between the two greatest pioneers of English Channel flight. There's a tinge of tragedy too in the knowledge that the woman in the photograph is about to pull off one of the great English Channel achievements; and that this remarkable individual, even at this distance exuding charisma, intelligence and oceans of untapped potential, has less than three months to live.

But while Quimby's flight was overshadowed by a marine disaster on the other side of the world, Blériot's provoked an immediate round of public soul-searching in England.

'A rather sinister significance will one day no doubt be found in the presence of our great fleet at Dover just at the very moment that, for the first time, a flying man passed over the "silver streak" and flitted far above the mast of the greatest battleship,' mused one newspaper the day after Blériot's flight. 'Of course such an event suggests all manner of fresh dangers as well as fresh advantages for the future, and it is one more evidence of some absurdity inherent in our civilisation that our first thought about aeroplanes would be about their use in war.'

'The arrival of Monsieur Blériot suggests to me how far behind we must be in all matters of ingenuity, device and mechanical contrivance,' sighed no less a scientific mind than H.G. Wells. 'In spite of our fleet this is no longer, in military terms, an inaccessible island.'

He then went on to predict that within a year aeroplanes would be crossing the Channel, circling London, able to drop explosives on the capital, and there wasn't a thing we could do about it. He was only four or five years out. 'What does it mean for us? One meaning, I think, stands out plainly enough, unpalatable to our national pride. This thing, from first to last, was made abroad.'

14

The Wreck of the *Amphitrite*

There are mornings when I'm out swimming and the English Channel seems like the most beautiful and benevolent thing on the planet. On particularly calm days when the air is so still I can almost feel myself brushing through it on my way down the beach, the surface of the water is so smooth I can watch the ripples I make spread out from me practically until they dissipate. On summer mornings when the sun rises and turns the water gold, even the shingle seems to glow. In winter there are mornings when the sky and sea are the same shade of grey-white and a light mist obscures the horizon, making it look as if the sky and the sea have become one; days when the surface is so smooth you could almost reach out and polish it. On those mornings it's almost as if time has stopped.

The English Channel may indeed be beautiful, especially on mornings like those, but benevolent it absolutely is not. The silky surface of that water hides centuries of storms, gales, disasters and tragedies. On the floor of this shallow sea lie millennia of wrecks and the bones of the countless. There are people who have slipped into these waters, by accident or design, never to be seen again: the suicides, the storm-tossed, the valiant, the naive, the adventurous, the timid, some of them not even missed. Many have turned up on Channel shores slumped in breakers or broken on rocks; they lie unnamed and unvisited in coastal churchyards, given dignity in burial by the kind. Most are still out there washing around on eternal tides, having long ago become part of the sea that took them.

Any investigation of our relationship with the Channel is soon subsumed in tragedy. Every location – town, village, city, port – on both sides has its memorial stones and plaques that can only hint at their stories: the worst storm in decades, lanterns by the shore, candles in windows, slow realisation, silence in a town, spaces in a generation.

The Channel has been inventive in its larceny, snatching every kind of vessel and taking people of all creeds, classes and colours. Swimming in the Channel is exhilarating and good for the soul, but it's also a firm reminder of the fragility of mortality. I make no claims to being a particularly strong swimmer, and the knowledge that going out a few strokes too far on the wrong kind of tide and I'll be joining the charnel of the Channel reminds me of the thin line between the joyous thrill of being in the cold water and the desperate fear of drowning.

One of the Channel's most successful deceptions is how harmless it appears to be. On our stretch of the coast you can even see the other side of it. Look at the map: it's a sliver between the dark waters of the North Sea and the vast Atlantic, two properly dangerous bodies of water. The Channel? People swim across it. Yet even just a few yards out from the shore, on mornings when the waves are higher than usual I can appreciate how easy it would be to be overwhelmed, as the wave crests lift me and drop me faster than is comfortable for my buoyancy and my head goes under. But ten or twenty strokes towards the shore, and my feet are touching shingle again. It is horribly easy to die in the Channel. Just as easy as in the deepest ocean or the tallest seas.

The more I found out about the Channel the more stories of death piled up around me. The more I swam the more I understood why, too. I didn't become inured to the countless tales of lost ships or desperate suicides, but certainly a level of repetition shifted the relentless accounts from bewildering to the almost commonplace. A few stood out, though, and none more so than the wreck of the *Amphitrite*, an enormous tragedy, one that was completely avoidable, and with some heart-rending individual stories.

At the tail end of August 1833 a hurricane-force wind blew across the Channel from the north-west for a full three days. Shipping came to a halt as craft pulled in to whatever shelter they could find. Nothing sailed in or out of Dover for two days. The *Ann and Amelia*, an East India Company vessel on her way back from India with a crew of thirty, was blown onto the French coast near Calais with the loss of four lives. The *Phoenix*, making the short hop from Deal to London, ended up on the Boulogne shore. The *William Friend* had made it across the Atlantic from Canada, heading for London, when it was wrecked on the shore at Calais; ten members of its crew drowned. In total a dozen ships were lost to that storm on the stretch of coast between Dunkirk and Boulogne.

In the days following the storm, once the Channel was passable again, news began to filter back of the dreadful loss of the *Amphitrite* along with 133 of her passengers and crew. Only three crew members survived, and the tale they brought ashore was one that horrified the nation.

Maria Hoskins must have believed there was more to life than scratching a living in London, or she would never have taken the watch. Twenty-eight years old, unmarried and living in St Martin-in-the-Fields, a stone's throw from where she was born, she lodged in Taylor's Buildings with a woman called Mary Haddon. On 13 February 1833 Haddon popped out for ten minutes and when she returned the watch she'd left hanging on the mantelpiece had gone and so had Maria Hoskins. Hoskins returned later and immediately told Haddon that she had taken it. Assuming from the ready confession that Hoskins had pawned the watch because she needed money urgently – she knew her lodger had been in and out of the workhouse and had recently been refused relief by the Covent Garden parish guardians – Haddon told her that if she handed over the pawnbroker's docket for the watch she would go and fetch it back and they would say no more about it on this occasion.

'No, I will not do that,' said Hoskins. 'I did it with the intention of being transported.'

Haddon pressed her further for the docket, to which Hoskins replied that she had destroyed it and would not reveal the name of the pawnbroker's until Haddon had called a policeman to have her arrested.

'If you have any compassion for a female you will take me up,' Hoskins implored her. 'If you do not, I shall do murder.'

Haddon had no choice but to accede if she wanted her watch back; she went out into the street and called over PC Broderick.

'I took the prisoner,' he said at Hoskins' trial. 'She said if she was not transported for this she would commit something more heinous that would guarantee her being sent out of the country. If she came across Mr Farmer she would drive a knife into him and hang for him.'

It wasn't the first time Hoskins had been in trouble. Three years earlier in May 1830, unable to pay a £5 fine, she had been sentenced to two months in prison for attacking Jeremiah Smith, the master of St Martin's Workhouse. Maria, it turned out, had a daughter Maria Ann, barely two years old in 1830, who she had been forced to place in the care of the workhouse. One afternoon when she went to visit her she found the child on the point of starvation. Understandably angry, she began shouting at the staff, attacking a porter and causing such a commotion that Smith himself arrived on the scene and grabbed her, intending to throw her out. Hoskins lashed out, cutting

Smith's eyeball and rendering him almost blind. She left with her daughter.

When she was convicted of the theft of Mary Haddon's watch and placed in Newgate prison until the time of her transportation, Maria Ann, now five years old, was back in the workhouse, but on 10 August she was returned to her mother at Newgate. Mother and daughter would be going to Australia together to escape whatever horrors and injustices had been bestowed upon them by the mysterious Mr Farmer.

The *Amphitrite* was a ninety-two-foot, three-masted ship built in north Devon and launched in 1802. She had regularly crossed the Atlantic with cargo in her early years before becoming a favourite charter of the Admiralty, bringing home soldiers from India and freed prisoners of war from Portugal. In 1833 she underwent a refit ahead of a new life: transporting convicts to New South Wales. On 25 August she set out from Woolwich destined for Port Jackson, Sydney, with a party of women convicts under the captaincy of thirty-three-year-old John Hunter, an Ayrshire man who had sailed the route before and who also had a significant shareholding in the *Amphitrite*.

It says much for the lot of the convict that nobody is quite certain how many women were on the *Amphitrite* when the ship set out along the Thames on a warm and sunny morning that betrayed no hint of the storms to come. There were more than 100, for certain, along with a dozen or so of their children. Some accounts set the number of women at 103, others at 108. Most had been in Newgate, having been brought to London from assizes all over the country to await their transportation. Maria Hoskins' story was just one of many, and while there were undoubtedly plenty of hardened criminals on the *Amphitrite*, some of the women had unbearably poignant tales of desperation caused by poverty.

From prostitution to pickpocketing, their crimes seemed to

attract arbitrary transportation terms: from seven years, like Maria Hoskins, to life. Eliza Smith was eighteen years old when she was sentenced to transportation for life for allegedly stealing a pocket watch from a man in Whitechapel. He'd claimed she'd rushed up to him, taken his watch and run off, but two witnesses said that they'd heard him tell Smith that he had no money and that she should take his watch to the pawnbroker nearby for a few coins. Smith's mother scoured the streets of Holborn for two weeks seeking witnesses to counter the man's accusation, but to no avail.

Thirty-six-year-old Charlotte Rogers was sentenced to fourteen years' transportation for stealing money from a man who had asked to go up to her room and stay the night. 'I am an unfortunate girl and cannot have a character,' she said in her defence, 'but I am perfectly innocent of the charge for which I am brought here.'

Martha Gates was nineteen when she was sentenced to seven years' transportation for stealing a shift, a petticoat and a shawl from her sister Maria. Maria told the court that her mistress, a Mrs Pounce of London, had banned Martha from her house and then noticed her in the street wearing her missing clothes. Pounce's son James later found a pile of Martha's clothes in the loft above the mangling room from where the stolen clothes had gone missing. 'We supposed that she got in there and had concealed herself for some time,' he said.

Mary Brown received a sentence of seven years for stealing thirty-three yards of cloth from the Coram Street, London, shop of William Owens. The nineteen year old presented a written defence in which she stated her husband Henry had been transported for life to Australia after being convicted of taking a handkerchief from the pocket of a Cambridge University student. 'I was left without means of support and in this state of distress I was induced to commit this crime with the hope that I might receive the like sentence of my husband and if possible be united to him again,' she wrote. 'I therefore humbly cast myself on the mercy of the court and trust that it will grant my request.'

Charlotte Smith and another Mary Brown, aged forty and twenty-six, had their death sentences commuted to transportation for life for stealing six shillings from John Charles Gates of Gray's Inn Road at the house in which both resided, labelled a 'house of ill-repute' by Gates. Smith said Gates had run into the building pursued by two men, so she called a policeman. Brown said she wasn't even there at the time.

Mary Hamilton was tried at the Old Bailey on 3 January 1833 for picking the pocket of a sailor named William Carter to the tune of a pocket watch and a bag containing twenty sovereigns. 'I cannot say the prisoner is the person,' conceded Carter at the trial, but Hamilton was sentenced to fourteen years' transportation anyway.

Prostitution and a bit of light larceny were about the measure of the women's crimes. They came to the *Amphitrite* mainly from London and Scotland, with a few sent from Worcester assizes and a number from Bristol, and all aged between sixteen and forty – some of them, the Scottish convicts in particular, repeat offenders. The convictions of some of the women charged as prostitutes with stealing from clients seemed particularly harsh. Most accounts have the men, most of whom agreed a price before going to the women's rooms, pleading surprise when they were charged. Whatever transpired in those small rooms in tenement buildings from Whitechapel to Dundee, it was the women's word against the men's, and in the 1830s, as often now, that only meant one outcome.

The separation from loved ones notwithstanding, transportation could provide opportunities for convicts they'd never have found in Britain, and there are numerous tales of transported men and women making a success of their lives in Australia. Some of them even returned wealthy.

Maria Hoskins' reasons for deliberately seeking to be transported can only be guessed at, but her life seemed to have been especially hard. In the year before her conviction she had endured three spells in St Martin's Workhouse, and her daughter, whom she'd found starved almost to death three years earlier, was still there at the time of her conviction. Maria Hoskins was desperate when she took that watch. One could even, reading between the lines, wonder whether the crime had been cooked up between her and Mary Haddon: the watch left hanging from a mantelpiece while its owner leaves the house for ten minutes seems quite convenient, and it would have been straightforward to retrieve from a pawn shop. From the start Hoskins was open about seeking to be transported, but unlike Mary Brown it seems she sought escape rather than a reconciliation. Abandoned by her husband William whom she'd married in 1823, her daughter nearly killed through neglect at St Martin's Workhouse, she was apparently wronged to such an extent by a man named Farmer that she felt transportation to the other side of the world the surest way of stopping herself from exacting a terrible revenge.

Either way, Hoskins clearly viewed Australia as a fresh start for her and her daughter, whatever the circumstances of their arrival in the country might be. While in Newgate waiting for the *Amphitrite* to depart, she wrote a letter to the county sheriffs that was reproduced in the newspapers after the tragedy:

> Being ordered for New South Wales, and entirely destitute of clothing, we, the unfortunate convicts in Newgate, humbly solicit a continuance of the Sheriffs' favour, that they will supply those that stand in need with clothing, as they have kindly done for other convicts. Some of us are entirely destitute, some are

more fortunate in being assisted by their friends. We are anxious to alter our way of living, and by strict adherence to the rules laid down for our future conduct, are in hopes partly to retrieve our reputation, which we, unfortunately, have forfeited. In mercy cast us not entirely from your favour, stretch out your hand to aid the repentant sinner, and your reward shall be great in heaven. Grant the prayer of your humble petitioner. On behalf of the convicts in Newgate, your very obedient humble servant, Maria Hoskins.

On 25 August 1833 these hundred-plus women, some of whom had been on board for weeks already, felt the *Amphitrite* finally slip her moorings and move into the middle of the river. They'd so far spent their time on board sewing, or learning to sew if they didn't already know how, thanks to the donations and tuition of a women's Quaker charity.

'Some of them appeared very well-disposed,' recalled a surviving crew member, John Owen, afterwards. Sarah Austin, who nursed Owen at Boulogne and took down his recollections, said she'd seen one of the women's boxes that had washed ashore opened at the Bureau de la Marine. 'The clothes were good and abundant,' she recalled. 'The small arrangements for future employment and housewifery, the little flat-iron, the neat cloth of needles, pins, cottons, and so on seemed adapted to something better than a life of disorder.'

Owen recalled the women were quiet and subdued, a few of them swapped bunks, and one young girl who spoke only Welsh sobbed quietly to herself. Whatever their circumstances, whatever the reason they had ended up here on a convict ship destined for the other side of the world, their next chapters had begun, and in the main were being greeted with quiet stoicism.

While Captain Hunter assumed command of the ship, the welfare and custody of the passengers were in the hands of surgeon James Forrester, who seemed to have a fairly laissez-faire attitude to the women in his charge. Whereas the hatches were battened down over them at night, for example, they were allowed to wander about the deck as much as they pleased during the day. His wife refused to have anything to do with the women, save for one that she co-opted as a personal maid. Forrester left them largely to their own devices and would only intervene if there was a serious breach of discipline. The only order the women received, according to Owen, was to carry their mattresses up to the deck every morning to air.

Having put in at Gravesend for supplies, the *Amphitrite* was off Dungeness on the 29th when the gale first blew up. Captain Hunter tried to battle the winds, but found himself at noon on the 31st off the coast of France around three miles east of Boulogne, where he ran aground. He managed to release the ship by setting the topsail and the main foresail, the gale lifting her off, but just outside Boulogne harbour, as the gale raged, the ship was grounded again, this time on a sandbank not far from the harbour, within sight of the town and around three-quarters of a mile from the shore. When Hunter let go the anchor, it became clear to the seamen watching from the shore as the storm grew fiercer that he intended to ride it out and wait for the tide to lift the vessel clear. The locals knew this was a calamitous idea – the rising sea would be sure to destroy a ship wedged as tightly as the *Amphitrite* obviously was.

A pilot boat was sent out to assist under a man named François Heuret, who hove to beneath the bows of the stricken vessel, informed the captain of his grievous situation and offered assistance in evacuating the ship. Captain Hunter refused his help and even prevented some crew members from going ashore with Heuret to help coordinate a rescue. Shortly after Heuret's return a local fisherman, Pierre Henin, took matters into his own hands, stripping off and plunging into the crashing seas in the teeth of the storm in order to swim out to the vessel.

In the meantime, according to one of the surviving crew members, there had been some discussion on board about lowering a boat. Ship's surgeon Forrester, a man who had treated Lord Byron in his day, and the captain seemed to have more of a grasp of how serious the situation was, but apparently Mrs Forrester put her foot down, first refusing to share a boat with the convict women and then suggesting that the officers and crew should save themselves and leave the women and children, still below deck beneath sealed hatches, to fend for themselves. Whatever the upshot of the discussion, the rescue boat was never used, John Owen recalling later that the argument ended with Forrester saying to his wife, 'Fine, in that case nobody is going in the boat.' The boat, said Owen, was capable of saving sixty people. Two trips to the shore would have saved everyone aboard.

It took Henin an hour to reach the *Amphitrite* in towering seas, an extraordinary achievement in itself.

'Give me a line to conduct you to land,' he called in English, 'or you are lost as the sea is coming in.'

Two lines were thrown to him, one from the stern, another from the bow. Henin reached the bow line and began swimming back to shore. Before long, however, he suddenly felt the rope jump from his grasp – it was being reeled back in to the *Amphitrite*. He swam back to the vessel and called for the line again, only to be refused. He later said he was sure the order had come from Hunter and Forrester that no line be sent out. Exhausted and not a little baffled, Henin swam back to the shore.

Shortly afterwards, around 7 p.m., the women below managed to force open one of the hatches and streamed onto the deck. Stunned to see land so close yet no moves being made to evacuate the ship, the women – possibly Maria Hoskins again to the fore – pleaded with Hunter to allow them into the boats

But met with stern and unequivocal refusal. The tide was rising and the *Amphitrite* was running out of time.

The witnesses on the shore reported the heart-rending sight of the women on board the ship, unable to save themselves yet unwilling to capitulate to the situation. Some estimated they were on deck for fully an hour and a half. The wind, callously, was coming from just the right direction for their screams to be carried clearly inland. The ship's timbers began to groan and the copper sheathing that encased the hull screeched in the gale as the rising tide fought the grounded vessel. Finally, with a loud cracking and wrenching the *Amphitrite* broke in two. The women on deck disappeared, cast into the water along with the crew and all the women and children who were still below.

John Owen clung to a spar with four other sailors and one of the women, and remained there for forty-five minutes before deciding to risk swimming for the shore. It took more than an hour for him to wash up half-drowned on the beach, to be hauled out of the surf and carried insensible to Barry's Hotel on the front. Another crew member, James Towsey, found himself in the sea clinging to a plank with another man. Asked who he was, the man replied, 'I am the captain.' When Towsey looked round again he had disappeared. The third survivor, a seaman named Rice, drifted ashore unconscious on a ladder.

It wasn't long until the bodies began to wash up: forty were piled up in the Humane Society's hut on the beach, which was equipped with just two beds and a washbasin

'I never saw so many fine and beautiful bodies,' said one correspondent in a report carried by many English newspapers. 'Some of the women were the most perfectly made.'

Once he had recovered enough to speak, Owen told the locals there were 136 people aboard: 108 women prisoners, twelve children and sixteen crew, but there are no official records to confirm this. The ship's manifest retained by the government lists 101 names, while a copy sent separately to Australia adds the name of Margaret Dunbar to that list, who boarded at Woolwich at the very last minute. From Elgin, she'd got drunk one night, stolen a horse belonging to a man named James Young and ridden it home, hiding it in a neighbour's barn. Newspaper reports of her trial, held only three months before the *Amphitrite* sailed, described her as a 'tidy, decent-looking woman' who had a number of excellent character references read out in her favour, but she was nonetheless sentenced to seven years' transportation.

Towsey, an old friend and shipmate of James Forrester, assigned all the blame to the captain. He should have driven hard for the shore and beached the vessel, he said. He fired no distress signal, hoisted no colours and displayed no lights. According to Rice and Owen, it was Forrester who persuaded the captain to reject Heuret's and Henin's offers of help. He was, it turned out, on a bonus of £1 for every convict who reached Australia safe and well, while the captain, it was reported, risked a fine of £1,000 for every convict that escaped from the *Amphitrite* while under his command.

'Who is to blame for all this?' asked a British resident of Boulogne who helped bring in the bodies washed ashore the morning after the disaster. 'The captain has been blamed for his obstinacy – but he is dead. The surgeon has been blamed for his tenacity – but he is lifeless. I have examined the three men who survive; all ought to have been saved. Where was the fault that they were not? We will have to wait and see.'

Further blame and accusations were bandied around in the weeks that followed. Even the British consul in Boulogne was in the firing line for not reaching the quayside quickly enough when it became clear it was a British ship in trouble, while letters appeared in the press blaming the French for not doing more to assist.

In the meantime, sixty-three dead women had washed up on the shore and the rest were still out in the water. Hunter and Forrester had had four hours in which they could have saved their charges but chose to do nothing, the captain seemingly hell bent on floating off the sandbank with the tide, even when the storm was still at its height.

'Sixty-three dead women were placed in long rows in a long room in the Hospice de St Louis in the Rue de l'Hôpital,' wrote another English resident of the town. 'Among them was a young mother with an infant still clutched in her rigid arms. Two or three hours earlier all of them were alive – all – and thought not to be even in danger, and now the half-naked and scarcely cold bodies were one inanimate mass.'

The bodies were placed in coffins and taken by cart to the English Cemetery in the east of the town. An English clergyman oversaw a short service and the sixty-three women were laid to rest in two rows. A few more bodies washed up in the days that followed and were added to the plot, creating a regular melancholy procession between shore and cemetery. Most of the bodies were naked, their flimsy clothes torn away by the sea, and even those who were clothed bore no clues as to their identities. They all died nameless.

The mass grave of the *Amphitrite* victims is close to the gate at what's now called the Eastern Cemetery on the outskirts of Boulogne. It's a quiet, shady spot next to the path, a sturdy plinth on a wide stone base with an urn shrouded by a cloth sculpted at its peak. The inscription on the plinth is only just discernible, battered as it is by almost two centuries of Channel weather. It reads: 'Here lie the bodies of 82 of the unfortunate persons wrecked on the coast at Boulogne on the English convict ship Amphitrite, the 31st August 1833'. The same inscription in French is engraved on the other side.

I imagine I'm one of very few visitors who've come here to

visit the victims of the *Amphitrite* during the two centuries since the ship went down. It's not a well-known disaster, most of the women would have left few if any descendants, and for those that did the tragedy is so long ago now as to be likely forgotten in family histories. They were nearly all from poor communities, so it's not as if their nearest and dearest could nip over to visit the grave anyway, especially those who would have come from Scotland. Even then, there's barely a third of the victims here; the rest are still out there somewhere, still part of the Channel that took them.

I'm glad the *Amphitrite* grave isn't hidden away at the back of the cemetery in some shady corner. It's among family plots and individual graves, lending the women an equality in death they never knew in life. Most of them never stood a chance from the start, shackled at birth by poverty and gender, used and abused, many of them having a price put on their bodies just to try and survive, but ultimately becoming victims of a terrible, preventable tragedy.

Their only dignity in death is the solemnity of the plinth under which they lie. I lay flowers I've brought from England at its base; when I step back I notice I've left the price on the cellophane. It's impossible to know if Maria Hoskins and her daughter are there. Was she the dead woman with the child still in her arms seen at the makeshift mortuary? Who knows. In a way I hope she isn't, that she's still out there somewhere amid the waters of the Channel, away from the poverty and hardship and injustice, eternally on her way to Australia towards a new life.

15

The Tunnel

Every time I'm sitting in the car encased in a plastic-lined train sarcophagus barrelling through the Channel Tunnel I can't help thinking about Aimé Thomé de Gamond.

'You know who would love to be in here right now?' I think to myself, drumming my fingers on the car steering wheel. 'Aimé Thomé de Gamond. Aimé Thomé de Gamond would really, really love this.' Then I notice my wife looking at me from the passenger seat and realise that I haven't thought this to myself; I've said it out loud.

'You and that flipping Aimé Thomé de Gamond,' she says. 'Every time we come through the Channel Tunnel it's Aimé Thomé de Gamond this, Aimé Thomé de Gamond that. It's like there's three of us in the car and one of them is flipping Aimé flipping Thomé flipping de flipping Gamond.'

There are plenty of flipping people I've come across in my Channel hopping, but few of them are as high in my affections as Aimé Thomé de Gamond. The English Channel draws a range of people towards it and the best of them by far are the dreamers. People like Blanchard, Blériot and Matthew Webb, names immortalised in Channel lore, but not just them, not only the pioneers and achievers, the crackpots and eccentrics – just anyone who has looked out from the Channel shore and thought, I wonder . . .

For some of us the boundaries of those dreams don't go

very far. I am entirely comfortable with my lack of desire to swim across the Channel, for example, but the sense of wonder and satisfaction I feel when the sun's coming up and I'm in the water and the day stretches out in front of me like the sea all the way to the horizon is plenty enough for me. But then there are the big dreamers, people like Aimé Thomé de Gamond, the ones who look out at the Channel and see possibilities unbound by logistics, practicalities and sometimes even reality, people who can look out at the sea stretching to the horizon and feel an innate certainty that it's going to shape their destiny.

For Aimé Thomé de Gamond his destiny – and he was absolutely certain of this for his entire life – lay not on the Channel or in the Channel but *under* the Channel. It's thanks to the vision, ambition and drive of Aimé Thomé de Gamond that we are able to cross the Channel by means of a tunnel today. The fact that I can startle my wife by suddenly talking about him when we're in the car on a train in a tunnel rattling through the Weald–Artois Anticline with thousands of tons of water, shipwrecks and sea life above us is in no small part down to Aimé Thomé de Gamond, who for more than three decades from the 1830s devoted his life to the creation of a tunnel under the Channel.

He wasn't the first to moot the plan as a good idea. As long ago as 1802 when the Peace of Amiens meant that Britain and France were, for once, not trying to shoot, stab, shell or drown each other, there was such giddiness at the prospect of mutual cooperation that a mining engineer named Albert Mathieu-Favier drew up plans for a tunnel bored beneath the Channel, and displayed them at the Palais du Luxembourg and the École Nationale Supérieure des Mines in Paris. He pictured a tunnel illuminated along its length by oil lamps, with a string of chimneys leading up to the surface for ventilation, a tunnel through which carriages and even pedestrians could easily pass, with an artificial island in the middle where horses could be changed and passengers could have a refreshing dose of sea air to break up their journey.

The plans eventually landed on the desk of Napoleon Bonaparte himself, who was so impressed that when he had a meeting with the British politician Charles James Fox he raised the subject and showed him the plans. Fox declared it 'one of the great enterprises we can now undertake together', but with Fox out of favour politically at home, not to mention Britain and France resuming trying to shoot, stab, shell and drown each other soon afterwards, the plans were soon forgotten.

Thirty years passed before the arrival on the tunnel scene of Aimé Thomé de Gamond. Born in Poitiers in 1807, de Gamond studied mining engineering in Brussels and the Netherlands, where he demonstrated an early penchant for going straight to the top by presenting King William I with ambitious plans to drain Lake Haarlem. Soon after qualifying, while working on a canal system for northern France, he began setting his mind to the issue of moving people across the Channel without getting (a) seasick or (b) wet. He was twenty-six years old in 1833 and already displaying an enormous bald head as well as the curious neck beard he favoured throughout his life when he came up with his first plan for a Channel tunnel. Unlike Mathieu-Favier, de Gamond went beyond mere drawings and took soundings of the seabed between Dover and Calais, enabling him to publish his first blueprint for a telescopic iron tube to be placed at the bottom of the Channel, in sections that would slot together and then be lined with bricks. The course of the tunnel along the seabed would have been levelled beforehand with a giant battering-ram operated from the surface by specially converted ships, with a giant rake to swish away the debris. It would, he estimated, take thirty years to complete; or fifteen, if both countries started building from their respective sides and met in the middle.

It was a plausible idea, but he was only just getting started. In short order followed plans to build a brick tunnel on the bottom of the sea, an enormous granite and steel bridge with arches higher than the dome of St Paul's Cathedral, and a vast floating concrete island that would be piloted between two jetties, each extending five miles out into the Channel. No idea was too wacky or out of bounds for de Gamond; at one point he even proposed a giant causeway from Dover to Calais over which traffic could drive, with shipping passing beneath three bridges incorporated into the design. This last plan he took to the Great Exhibition of 1851, where he aroused no small degree of interest until he revealed it would cost an estimated £33,000,000. He also encountered what he described as 'the obstinate resistance of mariners, who objected to their being obliged to ply their ships through the narrow channels', the spoilsports

After this setback, when he returned his focus to tunnels, he realised that soundings of the seabed weren't enough: what he needed was a proper sense of its geology. Even though the Channel is as shallow as 100 feet in the Strait of Dover nobody had any idea what was down there, but the person who would find out, thought Aimé Thomé de Gamond, was Aimé Thomé de Gamond. He would descend to the depths himself, to find out what was there.

What drove this single-minded dedication to opening up a direct link between France and Britain? Ideology. Not religious, although its adherents were pretty fervent in their beliefs, but the ideas of the French utopian socialist Claude-Henri de Rouvroy, the Count of Saint-Simon. Saint-Simon had been an active advocate of the French Revolution despite his nobility, and felt from his experiences and the mistakes made in 1789 that he could build a workable, prosperous utopian society in which everyone was equal. He theorised that human history alternated between periods of upheaval and periods of harmony: the French Revolution was the upheaval, and Saint-Simon proposed that the coming harmony be used to create an enduring society in which everyone had the same kind of opportunities no matter who or where they were. He set out plans for an aristocracy based on merit rather than on genealogy, overseeing a society in which social welfare was a priority and where the needs of the people trumped the needs of the individual.

De Gamond was, like many of France's restless intellectuals at the time, a committed Saint-Simonian, with a dream of a direct railway from London to Calcutta creating a moral union between East and West, an equality of peoples and sexes that could be explored by hopping on a train at Charing Cross and not getting off again until India, and vice versa. Tunnels through the Alps and tunnels through much of South America were also part of de Gamond's answer to world peace through international harmony, and when he started his geological experiments in the Channel this showed him clearly to be a man who could walk the walk as well as talk the talk.

Accompanied by his daughter, in 1855 the forty-eight year old took a boat out into the middle of the Channel between Dover and Calais, stripped naked, wrapped a roll of lint slathered with butter around his head and ears on the grounds that it would counter the water pressure at extreme depth, strapped on a belt from which bags of flints were suspended as ballast, secured a rope to himself, tied a red rope around his left arm that would act as a distress signal, and plunged in. Sinking to the bottom, he quickly scooped up some samples, released his belt ballast, then rose back up to the surface. As he grew more ambitious he would hold a spoonful of olive oil in his mouth as lubrication – believing doing so would allow him to expel air without the pressure at depth forcing water in – and strap ten inflated pigs' bladders around his waist to propel himself to the surface quicker once he'd detached his ballast. Despite the crackpot nature of the whole buttery, flinty, bladdery exercise, de Gamond succeeded in lowering himself more than 100 feet down to the bottom of the Channel and doing what he had to do, before ditching his ballast and rising by sheer bladder power back up to the top.

On one dive he found himself at the bottom of the sea being set upon by huge dogfish and conger eels, 'which seized me by the legs and arms. One of them bit me on the chin and would at the same time have attacked my throat if it had not been preserved by a thick handkerchief.' Ascending quickly, he found his assailants nipping at him all the way to the surface, where they were seen off by his oar-brandishing daughter.

To be fair to the marine life of the Channel, de Gamond, smeared with butter, spitting olive oil and surrounded by bobbing pigs' bladders, must have come across in the murk beneath the waves like some giant bacon sandwich, but even being attacked by the predators of the Channel didn't sway him from his destiny. The limited time he could spend on the seabed meant the samples he brought up were fairly random, but they were enough to convince him that boring a tunnel through the chalky bedrock between Britain and France was the best way forward. The plans for his giant Channel causeway were consigned to the bottom drawer of his bureau and fresh sheets of paper unrolled.

By 1856 he'd come up with his most detailed plan yet, a twenty-one-mile tunnel stretching from Cap Gris-Nez to East Wear Bay, just to the east of Folkestone below Capel-le-Ferne, where the Battle of Britain memorial is today. The tunnel would contain two railway lines and come to the surface at each end, nine miles inside both France and England, where the lines would connect with the existing rail networks of both countries, thus easing the passage of trains and passengers on their way to or from – if his grandest plans were to be realised – Calcutta. A series of twelve artificial islands would disguise the required ventilation shafts, which were also equipped with sea valves should either side wish to close off the tunnel by flooding in times of war or general disgruntlement.

At each end, in shafts about 300 feet below ground, would be the border stations for each nation. Halfway across, somewhere around the Varne sandbank, he planned a large shaft with a sizeable mid-Channel international port to be called the Étoile de Varne, featuring four quays, a lighthouse and living quarters for staff. It would be, he told a French newspaper, 'as spacious as the court of the Louvre' and provide passengers with somewhere to take the air, having spent a good fifteen minutes beneath the Channel. Indeed, so large was this shaft to be – it was originally intended as access for the building of the tunnel – that he even speculated whether the trains themselves might travel up to the surface and down again on tracks spiralling around its circumference.

The quays, he rhapsodised, would be 'garnished with the ships of all nations, some bound for the Baltic or the Mediterranean, others arriving from America or India'. Why these ships would want to anchor themselves to a random concrete island in the middle of the English Channel when they were destined for somewhere else entirely was a point not explored.

De Gamond was nothing if not persuasive, wearing down the staff of Emperor Napoleon III with his gushing enthusiasm until they granted him a meeting, at which the emperor told him he was so taken with the idea he would straight away form a scientific commission to investigate the plans further – 'as far as our present state of science allows' – and depending of course on the nature of future relations between France and Britain.

This prompted de Gamond to head across the Channel to persuade influential scientists and politicians that his plan was a solid-gold one – people like Robert Stephenson, Joseph Locke and Isambard Kingdom Brunel, who had not long completed with his father the tunnel under the Thames at Rotherhithe. His powers of persuasion granted him access to the highest offices in the land: at an audience with Prince Albert, de Gamond was so convincing that the prince went straight to the monarch herself and evinced a response from his notoriously seasick spouse that 'you may tell the French engineer that if he can accomplish it I will give him my blessing in my own name and in the name of all the ladies in England'.

A meeting with the prime minister Lord Palmerston didn't go quite so well. 'What?' erupted the crusty old septuagenarian when de Gamond smoothed out his plan across his desk and commenced pitching. 'You pretend to ask us to contribute to a work, the object of which is to shorten a distance we already find too short?' When Albert joined the meeting later in the day and expressed his enthusiasm, Palmerston grumbled, 'You would think quite differently, sir, if you had been born on this island.'

Palmerston's was a minority, if crucial, voice. The *Illustrated London News* mused that 'the realisation of this project would be attended by a double result of very great advantage to England', namely that 'it would still preserve to this country for the future that maritime isolation that formed its strength throughout the past', and that the communication and trade advantages would 'prevent the commercial isolation of which England would otherwise be threatened by the completion of the great railway networks which connect the centre of Europe without break to the ports at the east and west of the continent'.

The *Morning Chronicle* noted that many would consider the idea a lunatic one, but pointed out that the same people probably said the same thing about the railways, the electric telegraph and gas lamps. The *Northern Whig* came out strongly in favour too, enthusing about how 'the soils of England and France, long the homes of hereditary foemen, when locked together by the iron arms of an underground railway, may yet become the homes of hereditary friends, united heart and hand for the furtherance of human progress and civilisation'.

It was all heady stuff, which is why it's a shame that in the very same year that this heart-swelling talk of international cooperation looked like it might produce tangible results, politics started getting in the way again. An assassination attempt on Napoleon III by a bomb-lobbing Italian nationalist in Paris drew France into the Italian War of 1859, and then it emerged that the bomb used against the emperor had come from Birmingham. Britain responded as it usually does to matters European by wildly overreacting. Before the toys had even hit the ground, the talk of the nation was of an imminent invasion from France, and de Gamond's idea was pretty much dead in and, more significantly, under the water.

His combined faith in the human spirit and underwater engineering still burned brightly, however, and he was back in 1867 with a slightly revised version of his plans from a decade earlier. A certain amount of ennui had crept into his previously relentless conviction, though. Those 1867 plans would be his last. Thirty-five years of dedicated work had, he said, exhausted both his mind and his finances.

Three years later the Franco-Prussian War, in which France was overrun by spiky-helmeted forces from the east, dealt the fatal blow to de Gamond's plans, pushing talk of a Channel tunnel right into the background. The war also had a devastating effect on him personally. Already worn down by years of work, he was in Paris when the Prussians besieged the city, and then came the privations of the Paris Commune that followed.

In his final years his circumstances became so reduced that his daughter was working as a piano teacher to support him, until he died in his seventieth year on a cold winter's day in 1876 at his home on the Rue de Tivoli. He would have been delighted by the formation in 1872 of the Channel Tunnel Company Limited, for which he was invited to be consulting engineer, but by then he was too weak to contribute, the realisation that he wouldn't live to see the fulfilment of the dream to which he'd devoted his life contributing to the swiftness of his decline.

He would have seen, or at least heard about, one piece of tangible progress in the months before his death, however, when the company bought up land at St Margaret's Bay on the coast north-east of Dover for some experimental tunnel drilling. A French tunnel company was also formed, with financial backing from the Rothschilds, while a continuing political thaw brought the prospect of an actual tunnel back onto the agenda. But money troubles meant that test boring at St Margaret's never took place. Nevertheless, nearly 8,000 soundings were taken and 3,300 geological samples sourced and analysed. This was progress.

At this point the baton left by the death of de Gamond was picked up by Sir Edward Watkin, owner of the South-Eastern Railway company that ran the line between London, Folkestone and Dover. Watkin's first vision was of a direct line connecting his home city of Manchester with Dover and then continuing that line to the continent. He became the driving force of the tunnel project, the same kind of focus and face of the idea that de Gamond had been, even starting to dig a tunnel at Abbot's Cliff, midway between Dover and Folkestone, almost the exact spot de Gamond had identified as the site of his British tunnel station.

Watkin's project, with a new outfit called the Submarine Continental Railway Company, first saw a shaft sunk at the cliff site before the 1880s tunnelling works began to head out into the Channel as far as half a mile, followed by a second shaft sunk closer to Dover, at Shakespeare's Cliff. The success of these initial tests led him to plan further: a tunnel lit not by gas but by electricity, where trains were pulled by locomotives running on compressed air rather than steam. A station with a huge glass ceiling would be built beneath the East Cliff near Folkestone, while a shaft would bring the trains to the surface to connect with regular rail services to London by means of a vast hydraulic lift.

Watkin brought out the big bottles of oil to schmooze anybody who was anybody, staging sumptuous dinners beneath the Channel at which he entertained the likes of the Prince of Wales, the Archbishop of Canterbury and the Lord Mayor of London. But the tide of opinion was beginning to turn once more. Fear of invasion was growing in Britain again, even though the last tangible threat had ended the best part of a century earlier with the fall of Napoleon, and any attempt to bring the country into closer contact with the rest of the continent was generally greeted by the military top brass in the manner of someone having a paper bag burst next to their ear while enjoying a snooze. The Duke of Wellington had even been against the advent of the steam packet, worrying that the Channel was effectively being rendered an 'isthmus of steam', and he opposed the building of a railway line between London and Portsmouth on the grounds it would make it easy for an invading French force to storm the capital.

As well as an invasion threat that didn't really exist there was also the cost. While the tunnel was being funded entirely by private enterprise, Britain's geography of isolation meant that while its navy was regarded as the best in the world its standing army was, in terms of a major European power, small – around 60,000 in the 1870s compared to France's three-quarters of a million. The cost of the extra soldiers that would in the view of the military be necessary to respond to a column of French invaders popping up out of a single hole in the ground was too vast to countenance, and this was a view that proved enough to change Queen Victoria's mind on the issue.

'She hopes that the government will do nothing to encourage the proposed tunnel under the Channel,' the queen wrote to her prime minister Benjamin Disraeli in 1875, 'which she thinks very objectionable.'

By 1882 opposition to the tunnel was becoming as noisy as it was bonkers. The Archbishop of Canterbury got involved, heading a petition signed by more than 1,000 people fearing that the French reputation for free thinking might unleash the powers of the ungodly through any Channel tunnel. Cardinal Newman was against it too, on the perfectly normal grounds that 'Satan is unloosed on the streets of French cities; our turn may come next.' The petition was also signed by Robert Browning and Alfred, Lord Tennyson, the latter a particularly vocal opponent given his Francophobia and fondness for writing patriotic poems which seemed to involve English people with banners standing on hills and shouting a lot.

The adjutant-general of the Army Sir Garnet Wolseley, an old warhorse who'd seen the worst horrors of war in the Crimea and didn't fancy seeing anything similar on British soil any time soon, also reverted to patriotic terminology, harrumphing: 'Surely John Bull will not endanger his birthright, his property, in fact all that man can hold most dear, simply in order that men and women may cross to and fro between England and France without running the risk of seasickness?' He also helpfully composed a number of invasion scenarios that might have been useful to potential enemies, by which the crafty French in particular could render Britain defenceless by establishing a bridgehead at the tunnel entrance and funnelling troops and supplies through, calculating how many troops could reasonably fit on each train and how long they would need for the business of conquering Dover before marching on London. In 1888 Sir Randolph Churchill stood up in parliament to demand that 'our virgin isle retain her *virgo intacta*', which was a bit of an odd way of looking at it. The London offices of Watkin's (now renamed) Channel Tunnel Company even had their windows broken.

The French, utterly baffled by this English position, quietly pointed out they were 100 per cent in favour of the tunnel despite the fact that only one nation had sent invading armies across the Channel during the previous 800 years or so, and that was, well, England – roughly three times a century on average, not to mention hanging on to Calais for a good couple of hundred years.

A month after the archbishop's petition, the wording of which had included John of Gaunt's 'Sceptr'd Isle' speech from *Richard II*, the Board of Trade dropped a bombshell on Watkin, informing him they'd just remembered that the Crown owned the seabed, including everything under it, for three miles from the foreshore at low tide, so the work and the swanky dinners had to stop. Despite this wall of opposition that ranged from the vaguely rational to the paranoid to the outright bananas, Watkin never gave up, lobbying politicians hard and seeing a Channel Tunnel Bill presented to parliament in 1887 defeated by only seventy-six votes. By the following year he'd managed to convince the committed anti-tunneller William Gladstone to support another bill, but still it didn't quite pass.

He kept trying until 1894, when he decided instead to concentrate on developing a pleasure park at Wembley in north-west London, including an ambitious structure based on the Eiffel Tower. Construction reached the first stage, four vast feet and one platform, when it began to sink into the marshy ground and had to be abandoned. Known as Watkin's Folly, it was demolished early in the twentieth century. Wembley Stadium stands there now.

Watkin died in 1907, the same year that Edward VII, who as Prince of Wales had been in favour of the project after attending one of Watkin's legendary subterranean dinners, announced he was against the idea of a tunnel linking England and France. The king was a keen Francophile, spending three weeks of every year at Biarritz and at least a week in Paris, but somehow bracketed the tunnel with female suffrage as one of the modern world's potential abominations. His declaration was timed to coincide with another tunnel project, mooted this time by the French in light of the Entente Cordiale of 1904 and coordinated by Baron Frédéric Émile d'Erlanger. This plan was so obliging to Britain's obsession with being invaded that it offered to make the French end of the tunnel emerge from the cliffs at the small town of Wissant, visible to and in range of British coastal artillery should they ever feel the need to blow it to smithereens. Despite this generous proposal, another parliamentary bill failed to pass.

Two years later Blériot had flown across the Channel, ushering in a new age of communication, potential warfare and invasion paranoia. In the light of this, in 1913 d'Erlanger's son Émile tried a new approach, which failed in the Commons again.

The First World War, curiously, heralded a change of heart. It was noted how useful a tunnel would have been for moving troops and equipment quickly and without having to worry about sea conditions or the weather. In 1916 Arthur Conan Doyle wrote of how absurd it was that Britain was 'allowing ourselves to be frightened off from doing what was clearly to our advantage by the most absurd bogies, such as that we would be invaded through a rabbit burrow in the ground twenty-six miles long'. At the end of the war Marshal Ferdinand Foch, who had commanded the combined Allied Armies and orchestrated the Armistice, estimated that had there been a Channel tunnel the war could have been shortened by as much as two years.

Support for a tunnel grew through the 1920s – Winston Churchill, in contrast to his father, became one of the project's most vocal supporters, and when the great depression of 1929 hit, surprisingly the idea gained wider support as a method of creating much-needed jobs. A specially convened government committee was in favour, but the Imperial Defence Committee rejected it on entirely new grounds, now that the prevalence of aviation made potential subterranean invasion a non-issue. After much scratching of heads and chewing of pencils they succeeded in coming up with a one-two of brand new objections: the negative effect on commercial shipping and – and nobody saw this one coming – the risk of the tunnel causing ports to silt up.

The Second World War brought genuine fears the Germans might build their own tunnel, a theory nixed when they used the old tunnel shaft at the French end dating from Watkin's time as a rubbish dump, then landscaped it for use as part of a military cemetery. Support for the project grew apace after the war, with the defence secretary Harold Macmillan ruling out in 1955 the idea that a Channel tunnel could be opposed on military grounds.

Eventually, ten years later, after an agreement signed between Britain and France in 1964, work began from both ends, only for the British government to pull out unilaterally on 20 January 1975, ninety-nine years almost to the day after the death of Aimé Thomé de Gamond. The government cited uncertainties about Britain's fledgling membership of the European Economic Community, but mainly the fearsome recession in which the country was embroiled.

Finally, belatedly, magnificently, in 1988 work began in

earnest on two main tunnels and a service tunnel. In the latter, on 1 December 1990 a British tunnel worker, Graham Fagg from Dover, and his French counterpart, a Calais resident called Philippe Cozette, were selected as the men who would make the breakthrough and initiate the first cross-Channel contact made entirely on foot for about 8,000 years. With both countries carrying the moment live on television, Cozette in a beige boilersuit, safety hat and glasses picked up his drill and set about the flimsy wall, making a hole about two feet square at waist height. When the dust cleared there was Fagg's smiling face above his orange T-shirt, and the two men engaged in a firm handshake through the aperture of centuries. Plastic flags were exchanged, whipping frantically as the first ever subterranean cross-Channel wind barrelled through from the English side. Then Cozette and a colleague set about widening the hole through which Fagg could step, to cheering from both sides.

The Channel tunnel, shared by both the Eurostar trains and Le Shuttle, has been a roaring success. In 2017 10.3 million people travelled through on the Eurostar services, just topped by Le Shuttle which carried 10.4 million. In the same year 22,550,000 tonnes of freight also passed through, figures that would have seemed barely believable even to a rabid optimist like Aimé Thomé de Gamond.

While I find the Shuttle convenient and fast, for me the Eurostar is an experience that brims with Europe and all its possibilities. In its early years I lived within earshot as it passed through south-east London with a mournful, otherworldly moan.

Nothing sounds like the Eurostar, and few rail services give one such a frisson of European pizzazz simply by boarding a train. More than 150 million passengers have travelled by Eurostar since the first scheduled service left London's Waterloo Station at 8.23 a.m. on 15 November 1994, and despite its first passengers including Jeremy Beadle, Jeffrey Archer and the Kew Gardens Rotary Club, the Eurostar has seemed from the outset impossibly glamorous.

For the first generation of European Britons – those of us not old enough to remember a time before we were members of a European Union – the Eurostar reflected the European perspective of us Nineties twenty-somethings better than anything else. We had Britpop with its union flags, guitars and swagger, a whole new kind of national pride, a confident version that didn't have to base itself on superiority, the kind of inclusive national pride that chimed with other countries and that said this is what we have, we like it, we're proud of it, and we hope you like it too. It was a national pride based on solid self-esteem that came out of our more outward-facing sense of ourselves. That we could now hop on a train in London and disembark in Paris as easily as we could Manchester or Leeds seemed only natural, because the physical barrier of the English Channel was now merely theoretical.

I'd seen on television the moment on 1 December 1990 when Philippe Cozette and Graham Fagg reached through a small hole halfway under the Channel and shook hands in a gesture curiously reminiscent of the hands of friendship reaching across the Berlin Wall the previous year. That breaking down of the physical barrier between Britain and France made me almost as proud to be European as did the fall of the Berlin Wall, especially as Cozette and Fagg's momentous flesh-press came less than two months after the official reunification of Germany and a few weeks after Mikhail Gorbachev had been awarded the Nobel Peace Prize. This was Europe at its best: confident, democratic, its new generations able to put past conflicts behind them.

It was 1998 before I made my first trip on the Eurostar. Having extricated myself from both dole and Penge I was making occasional appearances on the *Big Breakfast* performing allegedly comic topical songs of questionable quality. During the 1998 World Cup I was dispatched to Paris in order to doorstep Desmond Lynam at his hotel and serenade him with a specially composed song hailing his peerless qualities as a broadcaster (understandably, Des was having none of this and I ended up on a street corner singing to a cardboard cut-out).

As far as breaking my Eurostar duck goes, wandering through Waterloo Station carrying a guitar case looking for two men carrying a TV camera and a lifesize cardboard Des Lynam was an unusual one, but what I remember most is the thrill of that first journey under the Channel. We'd eased out of Waterloo, shimmered across the South Downs and then everything went thrillingly dark.

Twenty minutes. That's all it took to emerge into the early evening sunshine among the villages, steeples and lengthening shadows of rural northern France. I'd spent most of the undersea period gawping at my own wide-eyed reflection in the window. It's the worst view in the world, yet it never fails to captivate me because there are hundreds of years of history distilled into those twenty minutes. Britain and France had been drawing closer together ever since Blanchard first crossed the Channel by hot-air balloon in 1785. When Louis Blériot crash-landed his plane just outside Dover one foggy morning in 1909, the gap between nations became even smaller.

It still took until the 1980s and an agreement between unlikely bedfellows Margaret Thatcher and François Mitterrand to see the project to its completion, even if Thatcher had strongly favoured a road tunnel ('I shall be the first to drive through the tunnel at the wheel of my car,' she'd said in 1985). The 2007 move to St Pancras only added to the sense of combined glamour and history: the sleek trains pulling in and out beneath the breathtaking iron latticework of William Barlow's vaulted Victorian engine shed and the thrill of riding the walkway up to the platforms from the undercroft departures area, originally a storage facility for barrels of Burton's beer. Best of all is the knowledge that, as you emerge next to your train, within a couple of hours or so you'll be in Paris, or Brussels, or Lille, or even Amsterdam – and with minimal fuss, as if crossing the English Channel were the most natural thing in the world.

The day I was passing through the tunnel on the Shuttle thinking about Aimé Thomé de Gamond was the twenty-fifth anniversary of the first services that ran on 6 May 1994. As part of the anniversary commemorations the media tracked down Graham Fagg, now retired and still living in Dover, who revealed that he had voted to leave the European Union in the 2016 referendum.

'I think that if you're going to stay in Europe you've got to stay much further in,' he'd told Euronews that morning. 'It would have to be a "federal states of Europe" where everybody is equal: equal tax, equal wages, equal benefits. But carrying on as we are just doesn't work.'

As I sat in the car, inside the train, inside the tunnel under the Channel, I couldn't help thinking that one person who would have been in favour of all of that, who would have been overcome at being inside my battered old car in a battered old shuttle train in the tunnel he dreamed would help bring continents together, was Aimé Thomé de Gamond. If I'd had any buttery lint with me I'd have wrapped my head in it in celebration.

16

Folkestone

The turn of the millennium wasn't exactly boom time for Folkestone. The opening of the Channel Tunnel had seen a decline in numbers using the east Kent ferry routes, while the end of duty-free shopping in 1999 saw the day-trip market diminish significantly. Following a blockade of French ports by fishermen protesting high fuel prices and the added competition from budget airlines, the last Folkestone–Boulogne ferry sailed in September 2000. During the 1970s 1.2 million people were passing through Folkestone harbour every year; by 2002 it was being used by ten fishing boats with a crew of around three each.

No regular passenger trains have arrived at Folkestone Harbour railway station since 2001 – bizarrely it was used by the *Orient Express* until 2009 – but a few maintenance and freight trains aside nothing has passed through since the dawn of the millennium. So when I say that of all my Channel travels nothing has evoked what I'm calling the spirit of the English Channel like Folkestone Harbour railway station, you'd be forgiven for thinking I'm at it.

At the end of a branch line that descends across a viaduct that splits Folkestone harbour in two, the station platforms sweep around to the left in a graceful curve, leaving the harbour arm at the far end just out of sight. With the departure of rail traffic from the harbour and no chance of Folkestone ever regaining its status as a major cross-Channel port, it would have been easy to demolish the station and the rest of the old port and give it over to the ubiquitous luxury apartment.

During the mid 2000s there were indeed those almost compulsory artists' impressions in the local papers, the ones that have little shrubby trees and stick-like people wandering around in them, pictures accompanied by mighty fine talk of seven-screen cinemas, casinos, restaurants and bars to go with all the luxury apartments. But no. Happily for Folkestone seafront the harbour arm has been given a £3.5 million facelift – one that doesn't involve fancy leisure developments, merely a sprucing up of the assets that are already there, including a renovation of the old station about which, I warn you, I am about to gush. It's beautifully done, reverential without being a pastiche, understated, clean and bright. The ironwork is painted a dark green, the walls and platform roofs are done in cream. The rails themselves are gone, but their echo remains in the platforms and rail beds comprising one big walkway sweeping out to the harbour arm, complete with flower boxes, seating and a couple of pieces of sculpture. It's airy, it's spacious, and best of all you walk all the way through the station without anyone trying to sell you anything.

Highlights, for me, are the signs. They're all in keeping with the colour scheme, white lettering on a green background, and they are a tribute to the golden age when Folkestone was a busy Channel conduit. 'Parcels, Left Luggage, Telegrams & Enquiries' reads one, with a French translation underneath. 'Ticket Office' reads another, accompanied by the French word *'Billets'*. I can't stop looking at these signs. I think it's the way both languages are written in the same font in letters of the same size, a small thing, but one that gladdens me as it seems to be a perfect example of the Channel as a connection rather than a separation, the two countries that border it fuzzing their boundaries at its fringes.

It's a sunny day when I pass through the station. A family with a kid in a buggy sits on the blocks of decking that are a cross between benches and steps down from the platforms to the track beds, and the parents are pointing out the flowers in a tub to a baby in its mother's arms who's showing not the slightest interest. Two teenagers displaying all the awkward body language of a first date sway awkwardly as they walk, both nervous, both frantically trying to think of something to say because they're at that fledgling stage of a relationship when conversational silence is terrifying rather than the reassurance of mutual ease it becomes. The only sound is the gentle breeze and the odd silence, but I'm so immersed in this place that I can hear the steam whistles of trains and ships and the pea-whistles of train guards. I can hear the rattle of luggage trolleys and sack barrows, the churn of paddle wheels, the cries of porters and the slamming of carriage doors. Folkestone Harbour station gives me space to imagine. It doesn't crowd me with information boards or interactive experiences. I'm left to fill in the gaps myself between the sweep of the platform and the tantalising hints of what used to be.

Out on the harbour arm itself there are picnic tables and food stalls, restaurants and diners in the old shelters, and a big blue double-decker bus dispensing quite delicious Greek food. Sitting at a bench inserting a chicken gyro wrap with extra halloumi into my face, the calm blue waters of the Channel alongside me and the start of the white cliffs shimmering into a Channel haze beyond, I don't think I could conjure a happier Channel moment than this. There aren't even any seagulls stalking the diners.

Completing the walk along the harbour arm I reach the lighthouse at the end, beautifully restored and converted into a bar, with no neon signage and no piped music, just the clean sandy-coloured brickwork and the phrase 'Weather is a third to place and time' painted on it in green. Walking back through the station, I circle the inner harbour and pass under one of the low arches of the viaduct to the small outer harbour, where I experience a shudder of recognition.

One of my earliest memories is of the Channel. It's a glorious sunny day and the water is sparkling. I'm with my mum and dad and we're walking towards Folkestone harbour to visit my grandfather on the boat he kept there for sea fishing. He died when I was five, so I'm very small, and Folkestone harbour looks enormous. As I remember it, my grandfather has just come out of the wheelhouse, caught sight of us, and is standing there in his dark trousers, braces holding them up over his white shirt, sleeves rolled to the elbows and his pipe between his teeth. He's wiping his hands on a rag that he drops to the deck, then waves, his Popeye forearms brown from the summer sun. His boat is a small cabin cruiser called the *Joshua Slocum*, named after the first man to sail single-handedly around the world, a vessel he'd bought in 1970 to replace his previous boat the *Yola*, which was moored close to his home at Greenwich until one night vandals got on board and sank it. Before my grandfather owned her, *Yola* had been one of the little ships at Dunkirk; my mum remembers seeing the signatures of the rescued soldiers inscribed on the wood panelling inside the cabin. As far as I know, *Yola* and her signatures are still somewhere at the bottom of the Thames.

Charlie White loved the sea and he loved fishing. He worked as a lorry driver in the London docks for most of his life, not to mention being a popular trade-union shop steward. Taking his boat out on the Channel was the perfect antidote to the grinding gears and relentless engine throb of the trucks. Today I have two of the pocket diaries he kept. The covers of both are covered in thick, black oil, so I have to keep them in polythene bags and every time I read them I'm left with a faint sticky residue on my fingers that's half a century old. They're not diaries in the Pepysian sense, more a record of where he was keeping the boat and how much it was costing him.

The earlier one is an *Angler's Mail* diary for 1970, small with a red plastic cover and the name of the magazine in gold on the front. The most significant entry in this diary was made on 3 May 1970, when he notes that he bought the *Joshua Slocum* for £460 from a man named Hoeffler living in Palmer's Green. The following weekend he moved his new purchase from the River Lea to the Thames at Greenwich, close to where he lived. Once there, he had to climb into the water to cut free a rope that had snagged around the propeller. It was an incident that, if the brief jottings in the diary are anything to go by, paved the way for an eventful relationship with his new vessel. As well as the snagged rope he blew two core plugs in the engine which needed replacing, not to mention a new anchor at a cost of £1, the old one having presumably been at the end of the rope he'd had to slice from around the propeller. On 1 June he paid £7 10s. for a berth at Newhaven, presumably where he intended to base the boat, before buying two fenders, a stove, three new Perspex windows, nails, screws and four jerrycans from Charlton marina.

The *Joshua Slocum* was proving to be an expensive hobby.

Leafing further through the thin tracing-paper pages of his diary, I learn that on Saturday 13 June 1970 Charlie set off from Greenwich for the English Channel at 7.30 a.m., stopping for the night at Herne Bay. He recorded that he'd covered eighty-eight miles but that the *Joshua Slocum* had been taking on water en route. The following day the rudder was swept away as he passed North Foreland, forcing an unscheduled stop at Ramsgate harbour. Two weeks and £27 on, a new rudder and new generator later, he departed for Folkestone. 'Weather bad,' he wrote.

The other diary I have is the *Motor Boat and Yachting Diary 1975*, its black cover also sticky with oil. On a page recording 'Details/Particulars of Vessel' I learn that the *Joshua Slocum* was thirty-two feet long with a beam of eight feet six inches and a draught of two feet six.

The first entry for 1975 is on 14 June when he writes, 'Out in boat fishing. Poor catch: 6 fish.' The following weekend, on 20 June, he 'stayed night in Vacation Hotel, Folkestone (£3.50), worked on boat all day'.

On 9 August things became a little more dramatic. 'Left Folkestone to fish 10.30am,' he notes. 'Weather blew up, anchor dragged. Fuel pipe bunged up, called for Dover lifeboat 12 o'clock, towed into Dover harbour, lost anchor and boat hook.' The next day he 'left Dover 11.30am, more petrol trouble on the way but made it safely, arrived Folkestone 12.30pm'.

On the 14th there's the first reference to non-maritime matters. 'Taken to Greenwich & District Hospital. Bad pains during night. Kept in.' On the 31st he notes, 'Birthday in hospital – 66', and the '66' is circled carefully: according to my mum, when he was a young man Charlie was told by a fortune teller that he would die at sixty-six. When he ringed that number in his diary he was seven months away from the heart attack that would kill him in his sleep on the night of 31 March 1976, sitting at home in his favourite armchair in front of the television, while my grandmother snoozed in her chair on the other side of the room.

Looking at these diaries, I think I can possibly fix the memory of my grandfather waving from his boat to a specific date: Sunday 22 June 1975, the day after he'd spent £3.50 on a bed for the night at Folkestone's Vacation Hotel. The previous day was my mum's birthday so there's a good chance that we climbed into

Dad's dark-blue Cortina for a birthday weekend outing to the Channel coast so Mum could see him. That day is my clearest memory of my grandfather. I can still see him now, waving at me, and still sense how excited I was about seeing him. When I emerge through the viaduct arch I realise I'm standing on the exact spot I was standing that summer's day in 1975. I can even pinpoint where the *Joshua Slocum* was moored – a large bright-red fishing boat is there now.

The harbour is busy that spring day, effectively the first properly warm day of the year. Small children run towards the ice-cream hut at the end as parents rummage in their pockets for change. The white cliffs continue into the distance beyond the harbour wall. There is a sudden splashing from the harbour, a joyously boisterous dog chasing a tennis ball down the concrete slope and into the water. A fishing boat converted to give pleasure trips chugs gently as the dozen or so passengers gather their things together and stand up ready to disembark, a couple of teenagers taking the opportunity for a last selfie.

A small boy in sandals and glasses, no more than seven, looks out across the water with a demeanour that is old beyond his years. Squinting against the sun and his hand in the pockets of his shorts, put a white beard on his chin and a clay pipe in his mouth and he could be a salty old seadog weighing up the tides and the weather at any point in the last 300 years.

Folkestone is a changing Channel town. It started out as a notorious centre for smuggling, and for a long period from the early eighteenth century practically ran on the proceeds of dodgy goods coming in from the Channel. When Daniel Defoe wandered through on the journey that made up his 1724 *Tour Through the Whole Island of Great Britain* he was astounded by the number of dragoons he encountered on the roads thereabouts, some of whom told him theirs was a thankless task that saw 'the wool carried off before their faces, not daring to meddle'. In 1777 an Act was passed forbidding anyone to loiter within five miles of the coast around Folkestone, and eventually in the 1820s a speedy customs brig named the *Pelter* was stationed permanently near the town, intended to battle the smugglers' vessels. Skirmishes and outright battles would take place out on the water and whole ships and their crews would disappear, never to be seen again. On moonless nights deadly tussles were fought between smuggling gangs and customs men, in coves, on beaches and on remote country lanes along the south coast and around Folkestone in particular.

This propensity of Folkestone folk to tweak the nose of both the law and convention extended beyond smuggling and into the fabric of everyday life. A Quaker named Jenkins, for example, who moved to Folkestone in 1820 and left some terrific journals of his impressions of the town, related how Folkestone men were always reluctant to marry unless there was a child already on the way. So rife were shotgun weddings there that a parson named Langhorne, who ministered in the town for twenty years in the late eighteenth century, offered a teapot to any woman who came to him to be married while still a virgin. In nearly two decades of his Folkestone ministry this policy cost him exactly one teapot.

It took the coming of the railway to Folkestone in 1843 to signal a change in the town's fortunes. The South Eastern Railway company immediately started a cross-Channel service between Folkestone and Boulogne, and when the company proceeded to assist in the building of a railway line from Boulogne to Amiens it created a through route from London to Paris which, if the weather conditions in the Channel were right, could be covered in twelve hours – unimaginably fast for the times.

When the new Victoria Pier was opened in 1888 it proved an instant success – 7,000 promenaders clicked through the turnstiles on opening day alone – and set the seal on Folkestone having become a destination of some considerable swank, known throughout the land as Fashionable Folkestone. Its snootiness was lampooned in an 1898 music-hall song performed by Marie Lloyd that referred to it as a place where 'coster[monger]s seldom go' because the locals are 'a stuck up lot you know'.

That changed thanks to a local businessman and aspiring impresario, one Robert Forsyth, who took on the lease of the pier in the early 1900s and immediately set about changing the character and perception of the town itself. He put on wrestling and boxing matches in the pier theatre, variety acts and, for the first time in Folkestone, film screenings that anticipated the coming mass appeal of cinema, anything that would park bums on seats. Folkestone came to host the first international beauty contests ever staged in Britain, an entente cordiale of competitive pulchritude that despite its unashamed sexism was far preferable to the traditional method of asserting Anglo-French superiority: killing each other.

The first official Folkestone beauty contest was in 1907 and the contestants were all men, a garland of gorgeous geezers who one by one stuck their heads through a large ornate picture frame placed on an easel at the point where the stage curtains met. The winner, and Folkestone's first beauty champion, was a Sergeant W. Hodgetts of the 8th Hussars, a veteran of the Boer War, stationed at the nearby Hythe School of Musketry and 'a handsome fellow with pearly teeth and a fine moustache'.

The following year came a version for women that, in the spirit of *Jeux Sans Frontières*, comprised a kind of semi-final for British contestants ahead of an international final that would decide the overall winner. The Pier Pavilion was packed on 14 August 1908 for the final, one that pitted six English women against an American with 'a catch of the season walk', whatever that was, and a woman from Paris. The line-up was completed by 'two Boulogne fisher-lassies, blue smocked and white capped, a flashing-eyed gypsy from the Austrian Tirol, two girls from Canada and, last but not least, two Irish colleens'. The audience voted via a ballot paper, as a result of which East Molesey's own Nellie Jarman was declared the winner with 450 votes – a landslide at more than 100 ahead of her nearest challenger – to become Britain's first beauty queen.

Folkestone's beauty contests became an annual event followed throughout the country, with national newspapers covering the proceedings, but the First World War brought such frippery to a close. Folkestone played a key role, with around ten million people passing through, mostly soldiers, of course, but also some 800,000 Red Cross and nursing staff and later around 3,500 German prisoners of war. While most of the traffic through Folkestone was heading further south, there was some movement the other way, and when in August 1914 the first Belgian refugees arrived in the town in a pathetic flotilla of fishing boats and coal barges, the realities of the war from which Britain was still protected by the Channel hit Folkestone. So traumatised were the Belgians that some even refused to leave the boats when they arrived, staring glassy-eyed and terrified at the locals gathered on the piers to receive them. Eventually they filed onto the quayside, some just in the clothes they stood up in, others carrying what few possessions they'd managed to salvage when the war came thundering to their door.

By early September more than 20,000 refugees had arrived in Folkestone, and still they kept coming: eventually an estimated 65,000 Belgians spent at least part of the war in the town – which surpassed itself. Almost immediately, a committee was set up to arrange for every arrival to receive a medical check-up and to set up canteens that provided thousands of free meals every day. The people of the town opened up their homes and took in the traumatised, shops began putting up signs in French, and a special French-language newspaper was produced in the town, *Le Franco-Belge*. In July 1915 Folkestone made sure it marked Belgium's national Independence Day holiday with a special 'Belgian Day'.

After the war Albert, King of the Belgians, registered his appreciation of the town's reception for his countrymen in need. 'Folkestone had earned the admiration not only of the Belgians, but also of the whole world: yes, the whole civilised world knew how the town of Folkestone had received them with such cordiality which would never be forgotten,' he said.

The loss of the ferry service hit the town hard around the turn of the millennium, but today it's becoming vibrant again, showing its unerring propensity to constantly adjust and flourish. I ruminated on how the Belgians arriving in great numbers were unquestioningly treated so well here. Yet today along this Channel shore the ancient fear of invasion has been stoked enough to make a few desperate people trying to cross the Channel into as big a reason to be frightened as when Harold II lined up his soldiers along the coast in 1066 waiting for the Normans to arrive, or when the nation was so spooked by the prospect of the French launching an invasion that the Duke of Wellington opposed building a railway between Portsmouth and London because those crafty Frenchies would only use it to attack the capital.

The rending of garments in some quarters that greeted Blanchard's balloon, Blériot's aeroplane and even the early Channel swimmers came from a place of fear, the same kind of fear that has a British home secretary declaring a national emergency while others holler about calling in the Navy to sail frigates up and down the Channel coast, and all in response to a few dozen frightened people trying to make their way from the European mainland in dinghies. We've spent too much of our Channel history looking out from our shores and expecting the worst. After all, the only people who've launched major invasions across the silver streak have been, well, us.

The small boy continues looking out at the Channel through his old man's eyes until his mother calls him and he runs off towards her, babbling something incomprehensible and laughing. The sun is warm and I squint down at the spot in the harbour where the *Joshua Slocum* had been moored that day in 1975. Again I can see my grandfather drop the rag to the deck and wave, the sun glinting off his glasses, his smile lopsided thanks to the pipe between his teeth, and suddenly I really miss him.

Then I have another picture, not a genuine memory I'm sure, but with my eyes screwed up against the light I feel I can see the *Joshua Slocum* sailing out of the harbour, the cross of my grandfather's braces clearly visible on the back of his white shirt as he stands at the wheel giving off the occasional puff of pipe smoke, seagulls milling overhead as the boat sails into the Channel haze towards the horizon at the centre of an ever-widening wake.

17

The Soul of the Channel

It was one of those warm spring days in the countryside when you sense everything around you bursting into life. The sun was shining and the year was renewing itself; the buds on the trees, the grass, the soil in the fields, even the air thrummed with freshness and youthful energy. I was perched on a low wall by a lychgate in front of an old parish church at the very end of a long winding lane. Gravestones crusted with lichen leaned with the fatigue of centuries over grass speckled with plump morning dew. Lambs bleated in the next field. You couldn't conjure a more traditionally English rural scene than this: the church with its squat square Victorian tower adjoining a chancel and nave dating back to the twelfth century, placed among fields and hedgerows where spring lambs gambolled and crops readied themselves to push through the soil.

You don't arrive at St Mary's Church at Capel-le-Ferne on the chalk ridge between Dover and Folkestone by accident – not unless you're really spectacularly lost. It's a long way from anywhere, even the village that bears its name, so nobody comes here without a reason. I was here because I was looking for someone. I'd just found her, too, beneath one of the less assuming of the churchyard's memorials, close by the lychgate.

Ellen Tough's simple headstone had once been topped by a sturdy stone cross but somehow that had been sheared off and was now propped behind it. The inscription was weather-faded and hard to decipher among the lichen blooms, but after a good deal of squatting and squinting in front of most of the other headstones in the churchyard my eyes had adjusted and eventually I'd found what I'd come to see:

ELLEN TOUGH

PASSENGER IN

NORTHFLEET

DROWNED 22ND JANUARY 1873

Ellen's grave is separate from the rest and its headstone faces at a different angle, compounding the sense of loneliness that characterises her presence here in a place to which she had no connection. She left little impression on the world during her life – even on her own headstone her name is dwarfed by that of a ship – but I felt compelled to find her as she seemed to represent something more than the sum of her life.

The *Northfleet* was a three-masted cargo-and-passenger vessel of 1,000 tons built in 1853 and already a veteran of the England–Australia run, when on 13 January 1873 she set sail from Gravesend with a hold full of iron rails and 379 people on board destined for Hobart, Tasmania. The iron was for a new railway and nearly all the passengers were labourers emigrating to build it, many of them with their families.

From the start the voyage was hit by bad weather and the *Northfleet* spent much of the first week riding out storms. On 22 January in the face of screaming gales she anchored about three miles off Dungeness, set her mast lights and hunkered down for the night. Late in the evening, when most passengers were in their bunks, the watch on deck saw a large ship looming out of the darkness heading straight for them. Despite their attempts to raise the alarm the ship, a Spanish schooner called the *Murillo*, rammed the *Northfleet* amidships with a splintering of timber and screeching of wrenched metal. Astonishingly the *Murillo*, instead of checking the damage and assisting in any evacuation, extricated itself from the stricken *Northfleet* and steamed off into the night, continuing on its way to Lisbon (the master would later claim that the damage had seemed superficial).

The ship sank within half an hour. While eighty-six people were saved via the two *Northfleet* lifeboats and two nearby vessels attracted by the ship's distress signals, the remaining 293 perished, of whom the Channel's tides and currents would give up only six. On 29 January, a week after the disaster, a woman's body was found on a beach close to Capel-le-Ferne some thirty miles from where the *Northfleet* went down. She was taken to the nearby Royal Oak pub, washed and dressed in a white chemise, placed in a parish coffin, photographed, and buried unidentified at St Mary's the following day, her only mourner a representative of the vessel's owners.

Meanwhile, from around the country photographs and descriptions of people feared lost in the disaster were arriving in Dover and it was noted that the dead woman bore a strong resemblance to the mother in a family group with a father and young daughter. The clothes in which the woman had been found and the photograph taken at the Royal Oak were dispatched to the sender, who confirmed they belonged to thirty-eight-year-old Ellen Tough from London. The cost of the headstone I'd just found was met by the Mansion House Relief Fund, set up in the wake of the tragedy.

Sitting on the wall that morning I ran through all I'd managed to find out about Ellen Tough. It didn't amount to much. She was born Ellen McCarty in Bantry, County Cork, in 1834, and spent her childhood on the shore of Bantry Bay looking out to the Atlantic beyond. At some point, probably around the time of the famine in Ireland in the mid 1840s, she emigrated to London with her parents John and Mary, and at the turn of the 1860s was working as a cook in a Notting Hill boarding house. In January 1862 at the age of twenty-eight she married Edwin Tough, a builders' labourer seven years her junior, and two years after that their daughter Ellen, known as Jane, was born.

It wasn't easy being poor in Victorian Britain and Edwin Tough was among those trying to do something about it, for his own family and for people like them. By 1872 he was president of the 2,500-strong Builders' Labourers' Union, and in September that year was petitioning the Master Builders' Association for a 6d pay rise for his members, as 'provisions, clothing and house-rent have risen to such an extent that our income cannot meet our expenditure'. The well-supported strike that went with the petition clearly didn't trigger much of an improvement, because four months later the Toughs were boarding a ship bound for a new life in Tasmania.

For Ellen it was to have been her second emigration. Having fled famine in Ireland she was on the move again, attempting to escape the grinding poverty that had defined her life. Instead the English Channel left her here, in a place she never knew and where at first nobody even knew her name. Members of her and Edwin's families, who would have been resigned to never seeing her again after she'd emigrated – the family photograph from which she was identified was probably taken specially, intended as a keepsake – visited her grave three weeks after her death, but since then? There's every chance I was her first visitor in almost 150 years. The small bunch of flowers I left seemed pathetic.

Although no longer used for services, St Mary's is maintained by the Churches Conservation Trust and is open to visitors determined enough to find it. I slipped off the wall, walked into the porch, unlatched the heavy 800-year-old wooden door and stepped into the cool interior. As my eyes adjusted to the light, I noted immediately that atmosphere unique to old country churches, where centuries of quiet contemplation leave a meditative stillness that's tangible even to the non-religious. It's a simple church with dark wooden pews and whitewashed walls discoloured in patches by damp, but one by which I found myself instantly charmed. The wartime roll of honour is handwritten, hangs in a frame and runs to just four names. The chancel is separated from the nave by a beautiful three-bay carved stone arcade from the fourteenth century that has two crude faces carved into it at eye level, from which the eyes are drawn to the beamed roof supported by timber that's almost a millennium old.

I stood at the spot where Ellen's coffin would have rested during her brief funeral service. The vicar's voice would have echoed around a church empty save for the company man standing awkwardly in a pew with his hat in his hand, there for what he represented rather than who he was and keen for this all to be over as soon as possible. I looked down at the stone floor of the aisle where Ellen's coffin would have stood, and saw an old ledger slab on which was inscribed a name I recognised even though the lettering had been worn by centuries of footsteps. I looked up and saw the same name on a wall-mounted memorial plaque, and then another. There are Minets commemorated here: Alice, who died in 1778 and lies beneath the aisle, and on the plaques Mary (1768) and another Alice (1855). There's also an elaborate plaque to General Sir Charles Staveley, a distinguished Boulogne-born soldier who served in the Crimean War as well as in China and Abyssinia, and who married the daughter of Charles Minet.

The influence of this Dover-based family stretched far and wide for a good 200 years and beyond, ranging from this small, out-of-the-way church to one of Britain's biggest banks.

The Minets represent one of the country's greatest refugee stories.

Early in the morning of 1 August 1686, a small gathering watched a boat arrive at the beach by Dover harbour and its occupants drop into the shallows and wade onto the shingle. Above the roar of the waves and the gulls, cries of joy and relief sounded from both groups for the ones who'd just stepped off the boat; for the first time in many months there was a feeling that they had reached a place of safety.

The previous year the Catholic King of France, Louis XIV, had outlawed Protestantism by revoking the century-old Edict of Nantes, leaving the nation's Protestants essentially facing a choice between converting and fleeing the country. Thanks in part to a virulently sectarian local bishop, anti-Protestant feeling was particularly venomous in Calais, where there lived a grocer named Isaac Minet. The Minets were a long-established Calais family, prosperous, popular and Protestant, but the forces unleashed by Louis XIV saw them victimised and abused until they could finally escape across the Channel after months of degradation.

When he landed that summer morning in 1686, the twenty-six-year-old Isaac was one of around 50,000 Huguenot refugees who found sanctuary on the other side of the Channel in the wake of Louis's crackdown. As he squelched ashore Minet would have been familiar with the town that lay before him. His father had sent him, at the age of fourteen, from the family home on Calais's Place d'Armes to Dover in order to learn English and the intricacies of cross-Channel trade: Ambrose Minet had a thriving business in groceries, liquor and tobacco and was already lining up his teenage son as his successor. While

Ambrose knew that cross-Channel trade was the cornerstone of his business, he could have had no inkling of how the English town would come to define the family legacy. Isaac returned home two years later fluent in English and imbued with Dovorian commercial practice, so when Ambrose died in 1679 the nineteen year old was more than ready to take over.

In 1685, almost as soon as the ink was dry on Louis's revocation, anti-Protestant feeling fell upon the Minets in grisly fashion. Isaac's mother Suzanne came from a leading Calais Protestant family too, and when her sister died in the late summer of that year she refused the Catholic sacraments on her deathbed. Such was the local outrage at this news that her body was taken to the town prison and her estate confiscated. Three days later her corpse was dragged through the streets, where crowds abused it to the point of decapitation, then it was nailed to a wooden stake on the outskirts of the town. Isaac and his mother were put under house arrest and two guards were billeted with them. After a few days they learned a troop of dragoons was preparing to come and question them, something that could not end well for the Minets. Thinking quickly, Isaac succeeded in getting the guards roaring drunk, and as they snored boozily by the fire he and Suzanne managed to escape to the home of a friend, a Dutch chandler named Fournier who was prepared to risk sheltering them.

So keen were the authorities on tracking down the Minets that a 100-*livre* reward was offered for their capture, and a 1,000-*livre* fine threatened on anyone found to be harbouring them. They hid for four days during which, with Fournier becoming increasingly anxious, Isaac realised their only chance lay in escaping the town and making their way across the Channel. Word was sent to a man who had a house on the beach outside the city walls that he would be well rewarded if he was prepared to hide them, something to which he agreed readily when he heard how much was in it for him. Suzanne went first, disguised as a maid and carrying a water pail, then Isaac followed later dressed as a blacksmith. When he arrived at the house it was to the news that his mother had been recognised en route, arrested and taken to prison.

Believing he had a better chance of securing his mother's release from Protestant England than from a draughty hayloft outside Calais, Isaac wrote to his brothers, already in England, asking them to send a boat. A date was set but the householder went into town one night, got drunk, became notably gener-ous with his suddenly bulging purse and ended up blabbing the whole plan to fellow drinkers. Seeing soldiers crossing the sandy scrubland towards his hideout, Isaac hid in a haystack, was discovered and joined his mother in the prison, where, as he later recalled, 'we laid on dung in a very stinking hole'.

Six weeks of relentless harassment to recant their faith fol-lowed. The prison's superintendent, Isaac later recalled, 'told me I was a heretick & smelt strong of faggots and that I should be burnt etc' until he and Suzanne finally agreed to renounce their Protestantism in exchange for their freedom. When they returned to their home they found it had been ransacked and everything of value taken.

More determined than ever to reach the sanctuary of Eng-land, Isaac arranged a passage along with his mother, sister and her husband. Crossing the Channel in the seventeenth century was a hazardous undertaking at the best of times, but on top of changeable weather, tides and sea conditions the French authorities had soldiers patrolling the coast and cutters in the Channel hunting down Huguenot refugee boats heading for England. A successful escape would take ingenuity, courage and a healthy dose of good fortune.

The rendezvous for the Minet party was a remote house on the road from Calais to Gravelines on the night of 31 July, but at the appointed time there was no sign of Isaac's sister Elizabeth and her husband. Aware that the boat's arrival was imminent and the tides allowed little room for delay, Isaac borrowed a horse and rode back towards Calais, where he found the missing pair on the road trying to fix a broken wheel on the cart they had hired. When the trio finally reached the boat they found a farmer, his wife and their six children already sitting in it: somehow they had heard about the plan and begged to be taken across to England. Isaac eventually agreed to let them remain in the boat, but the boatmen warned them that if the weather turned and the boat was in danger of being swamped, farmer and family would be cast into the water with their belongings in order to save it. The fact they readily agreed to this demonstrates the level of persecution faced by people of their faith. The risk of drowning was one they were fully prepared to take.

The boat slipped away from the beach in the early hours and – one hairy moment aside when a Dunkirk cutter sailed frighteningly close and the boatmen had to throw a spare sail over their passengers – the crossing was made without mishap. When the Minets waded ashore and recognised the people waiting for them they were overcome, 'full of tears of joy in our eyes and many more in those of our friends who received us as brethren saved from the great persecution'.

Having connections in the town already, unlike most of the Huguenots who put ashore at Dover the Minets stayed put. Isaac established a shipping company that was so successful he would become a freeman of the town, and by the 1720s was able to build a mansion overlooking the harbour. It's long gone but it stood next to the modern main road running along the front from the eastern docks, where hundreds of cars and lorries now thunder past every day on their way to and from Calais.

So much money did the Minets make – a hefty portion of it from smuggling and a form of legal piracy that thrived during the War of the Austrian Succession – that in the early 1740s they were able to set up their own bank. By the time Isaac died in 1745 at the age of eighty-four he had passed the reins on to his son William and seen him already increase the family's fortunes. When William joined forces with Peter Fector, Isaac's nephew and a Dutch immigrant who had arrived in Dover at the age of sixteen, the enterprise grew even larger, becoming iron-clad when Fector married into the Minet family to create one of the most important dynasties in the history of Dover. The town practically ran on Minet money and influence, and eighty years after Isaac's dramatic arrival his grandson Hughes Minet became Dover's mayor.

On Peter's death in 1815, his son John Minet Fector took over the family businesses and made such an impression he provided the inspiration for Charles Darnay in Dickens' *A Tale of Two Cities*. A popular man around the town (unlike his father Peter, who grew bitter when he was removed from the town council when it was confirmed he was born abroad), John married the daughter of a Scottish MP, completing the family's rise from frightened refugees washing up on the coast to vital and popular pillars of the local community at the heart of the British establishment. The bank the Minets founded in Dover went on to become the National Westminster Bank.

You can't see the English Channel from St Mary's at Capel-le-Ferne. You can't even hear it. There's a good mile or so between the church and the cliffs, where at the Battle of Britain memorial a giant stone airman sits cross-legged looking out to sea. Yet standing in the stillness and silence of that ancient Norman chapel-in-the-ferns that gave the village its French name, without the sound of the waves, without immersing myself at dawn in its chilly waters, without giant container ships like silhouette Manhattans in the distance, without ferries continually plough-ing there and back, and without the orange sodium glow of French streetlights on the horizon, I felt like I'd found some-thing approaching the soul of the English Channel. Here was its tragedy in the truncated gravestone of a poor emigrant, and its benevolence in the memorials to a family who made good and that it led away from danger.

It could have been different; all it would have taken was a stormier night at the end of the seventeenth century and a slight alteration to the course of a Spanish schooner a couple of hun-dred years later, and those outcomes might have been reversed. The same is true of other Channel days: if the wind hadn't blown as it did in the face of the Armada, or if it had continued to blow in the late summer of 1066 and kept William at bay for the winter, or if the White Ship had waited until morning, or if the Channel hadn't been glass-smooth in June 1940 . . . A tide, a storm; our history and identity have persistently turned on a Channel whim.

I lay awake that night somehow more aware than ever of the ebb and flow of the Channel barely a hundred yards from my bed. I thought of the people I'd encountered over many centuries during my Channel winter, so by the time the light began to seep around the blinds my mind thrummed with Jean Bart, Charlotte Turner Smith, Jabez Wolffe, Mercedes Gleitze and the rest. I thought about how in 1849 Herman Melville had come ashore outside my window, disembarking from the steamer *Southampton* from New York at the spot where 'some centuries ago a person called Julius Caesar jumped ashore about in this place and took possession'.

Several hours later there were a few high clouds tinged with the pink of the pre-dawn as I crunched across the shingle. The tide was high, the water dark and calm, and on the horizon the hump of France seemed to be floating slightly above the surface in the soupy morning air. I put down my mug of tea, removed my robe, waded into the water, lifted my feet from the bottom and began to swim. As I did so the sun appeared, lifting itself slowly over the meniscus of the horizon and throwing golden shards across the surface of the deep-blue water. I swam towards it, stroking through the shifting, glistening surface that brims with the stories of the Channel people.

Author's Note

I remove my cap and mash it in my hands in bashful gratitude to Lizzy Kremer, Alan Samson, Lucinda McNeile, Sue Phillpott, Paul Murphy and Elizabeth Allen, without whom this book could not have happened and instead all the stuff in it built up in my head until it became so heavy I'd be unable to lift it. Phew.

Rather than list all the sources I consulted, here are some that I found particularly useful and enjoyed reading. Among general histories there were *The English Channel* by Nigel Calder, *The English Channel: A History* by J.A. Williamson, *The Narrow Sea* by Peter Unwin and *The Channel* by Shirley Harrison. *Gone to the Continent: The British in Calais 1760–1860* by Martin Brayne and *Sixty Miles From England: The English at Dieppe 1814–1914* by Simona Pakenham were invaluable and very readable accounts of those aspects of the two towns. Lorraine Fletcher's *Charlotte Smith: A Critical Biography* provides a scholarly yet accessible insight into that remarkable woman. *The Tunnel Under the Channel* by Thomas Whiteside was an absorbing account of an entertaining saga, while *In the Wake of Mercedes Gleitze, Open Water Swimming Pioneer* by her daughter Doloranda Pember and *The Crossing: The Curious Story of the First Man to Swim the English Channel* by Kathy Watson were both extremely enlightening about two extraordinary swimmers. Special mention too for the *Text-Book of Swimming*

by Jabez Wolffe, in which arguably the most entertaining character ever associated with the Channel teaches you to swim. By
far my favourite title was *Her Mentor Was an Albatross,* with
which Henry M. Holden christened his biography of Harriet
Quimby in magnificent style.

For more tales of the sea do seek out my podcast 'Coastal
Stories'. You can find me on Twitter @charlieconnelly and I
post pics almost daily from in and beside the Channel on Insta-
gram, @chasconnelly.

As ever all my love and thanks go to Jude, who is the best.